Minding Time: A Philosophical and Theoretical Approach to the Psychology of Time

Supplements to

The Study of Time

VOLUME 5

The titles published in this series are listed at brill.com/sst

Minding Time: A Philosophical and Theoretical Approach to the Psychology of Time

By

Carlos Montemayor

BRILL

LEIDEN · BOSTON
2013

Cover illustration: Aphelocoma woodhouseii (Western Scrub Jay), Santa Fe, New Mexico. Photograph: Peter Wallack. Wikimedia Commons.

Library of Congress Cataloging-in-Publication Data

Montemayor, Carlos.
Minding time : a philosophical and theoretical approach to the psychology of time / by Carlos Montemayor.
p. cm. -- (Supplements to the study of time, v. 5)
Includes bibliographical references and index.
ISBN 978-90-04-22891-7 (alk. paper) -- ISBN 978-90-04-23617-2 (e-book) 1. Time. I. Title.

BD638.M635 2013
115--dc23

2012029904

This publication has been typeset in the multilingual "Brill" typeface. With over 5,100 characters covering Latin, IPA, Greek, and Cyrillic, this typeface is especially suitable for use in the humanities. For more information, please see www.brill.com/brill-typeface.

ISSN 1873-7463
ISBN 978 90 04 22891 7 (hardback)
ISBN 978 90 04 23617 2 (e-book)

This book is printed on acid-free paper.

MIX
Paper from responsible sources
FSC® C004472
FSC www.fsc.org

PRINTED BY DRUKKERIJ WILCO B.V. - AMERSFOORT, THE NETHERLANDS

To Julius Thomas Fraser
In memoriam

CONTENTS

ACKNOWLEDGEMENTS

I owe the deepest debt of gratitude to Alvin Goldman, whose careful guidance was crucial for the completion of this book, which is a revised version of my dissertation. I am also very thankful to Frankie Egan and Dean Zimmerman for suggestions that led to critical improvements. Likewise, I wish to thank Randy Gallistel, whose work inspired many of the proposals I offer, and who was always ready to help me with clarifications and comments.

I first started thinking about the topic of this book because of conversations I had frequently with Zenon Pylyshyn, whose work has heavily influenced the way in which I think about issues in the philosophy of mind. My interest in timing mechanisms and metric representation greatly increased because of conversations with Rochel Gelman, Fuat Balci and Harry Haladjian. I am very grateful to all of them. Thanks to Jerry Fodor, Barry Loewer, Bob Matthews and Brian McLaughlin for helpful advice. I also benefited from lively conversations with the members of the International Society for the Study of Time that participated in its 2007 conference at Asilomar.

Many thanks also to Victoria, my mother and my sister for their love and unconditional support.

This book is dedicated to J. T. Fraser, who encouraged me to develop my ideas about time. His interdisciplinary work on time, as well as his generosity and kindness were crucial in revising this book.

PREFACE

Suppose there is no time. There is only a frozen grid that scientists call spacetime, but nothing that corresponds to our experience of time. Every single event in the history of the universe is somehow there in the frozen lattice: your birth and death, as well as the birth and death of our galaxy. Nothing really changes and time does not pass: the frozen lattice is all there is. It seems very hard to imagine this, but scientists believe it is true. A very important reason for finding it so hard to imagine is that we cannot conceive of a single experience in which time is not passing. The passage of time seems to be a fundamental aspect of anything we can possibly experience. Without it, nothing about our experiences as conscious beings makes sense.

According to some scientists, we should really try to make sense of the possibility that we inhabit a timeless frozen grid because there seems to be no room for what we normally call time in physics. The equations of physics are time reversible; there is no unique moment in time that could correspond to the present, and there is no distinction between past and future. The passage of time in your life, as you change and grow older is, in the light of the equations of physics, just an illusion. This is indeed a very hard truth to swallow. This book seeks to clarify several issues with respect to the puzzle of why we represent and experience time, without denying that physics has the last word on the nature of spacetime. It is beyond the scope of this book to establish what time is.

The main question of this book is: even assuming that time is an illusion, *what could create such a powerful illusion*? Our awareness of time permeates our lives. It is so familiar that we take it for granted. It is the background for our decisions and actions. It frames our emotions, struggles and experiences. Our awareness of time is, metaphorically speaking, a constant companion. For that reason, it is a very pressing issue to understand the causes of the experience of the passage of time.

With the development of ever more precise clocks and techniques for measuring time we have become much more reliable at representing and calculating time. Technology has had a significant impact on the way in which time dictates the rhythms of our lives. Scientific and cultural representations of time are more complex than ever before and make possible the intricate web of synchronized interactions that are characteristic of our age.

But how exactly does the *mind* represent time? Answering this question is crucial regardless of issues concerning the existence of time. How should we characterize our awareness of time, which is the origin of other, more complex representations of time? Until very recently, these questions needed to be answered only by intuitive insight and philosophical analysis. Philosophers offered various theses about how the mind represents time, but no evidence in support of these theses was available.

The psychology of time—particularly the study of how the mind estimates the duration of intervals and is acquainted with the present—has become a very important area of research in cognitive science. Experiments with animals that are very precise in time-sensitive tasks have produced valuable evidence, which shows that there are mechanisms for *representing* time (cognitive clocks). The representations produced by these mechanisms have a structure that allows these animals to combine them with other representations, such as spatial representations and even episodic-like memories.

What is a mental representation of time? Throughout this book, I shall assume that representations of time, like any other mental representation, have content and non-trivial veridicality conditions. A mental representation is about something (an object, an event, or a fact), and it may successfully capture reality. This is why veridicality conditions are important. My thoughts about the Minotaur are false, my beliefs about string theory may be false, and my belief that the Eiffel tower is in Paris is true. Although the role of veridicality conditions in perception is a subject of debate, I shall assume that veridicality conditions distinguish perception (for instance of color) from illusion and hallucination.

The representations of time I will analyze are best understood as representations produced by our sensorial apparatus. They have veridicality conditions and, therefore, may be considered to be types of perceptual representations (although veridicality conditions are unique in the case of time, as I am about to explain). Despite some problematic issues, mainly concerning the lack of specific stimuli for time perception raised by J. J. Gibson (1975), I will argue that such problems may be partly addressed by a careful assessment of the constraints on *any accurate measurement of time*, including scientific representations of time. My proposal is inspired by the work of Hans Reichenbach (1958), which remains one of the best philosophical studies of the constraints on spatiotemporal measurements.

I shall also assume that the content of representations is constituted (at least partly) by their veridicality conditions, in a way in which such content is susceptible of cognitive integration. I discuss this in detail in chapter 3, but the main idea is that representations of phases of a time cycle or segments of an interval can be *attributed* to a particular event or category, which allows animals and humans to structure their behavior temporally (See Burge, T., 2010, for the importance of this constraint). Chapters 2 and 3 show that all these constraints are satisfied by temporal representations in animal cognition.

An important feature of my analysis of temporal representation in chapters 2 and 3 is that it demonstrates that these representations are legitimate mental representations with content, without assuming controversial views about the self, conceptual content or causality. In this sense, my approach differs from, for instance, John Campbell's (1995) analysis of circadian and interval clocks in animals. The focus of my account is temporal representation of the most fundamental nature, without requirements of causality or the self. Some findings suggest that episodic memory is possible even in the absence of these representations (particularly the findings on scrub jays that are discussed in chapter 3). So it seems adequate to address exclusively temporal representation without introducing representations about the self and causality.

Chapter 4 presents an innovative proposal concerning the psychological evidence on simultaneity windows and the present moment. Its main advantage is that it allows for the present moment to be a kind of sensorial marker that satisfies the third constraint on any coordinative definition of time (i.e., simultaneity), while also allowing for the possibility of a different type of present moment that is associated with the phenomenal character of subjective experiences. These two different types of present moments are crucial components of our psychology, and there is evidence for both of them. One is required by the constraints on time measurements and the other by our conscious experiences. However, no account has addressed the relation between them, based on the evidence. Another advantage of this model is that it clarifies why the current models for temporal consciousness may not be incompatible, because they may address only one of these two different issues.

The scientific findings on which this book focuses concern two biological cognitive clocks (the circadian clock and the stopwatch) and what psychologists call 'simultaneity windows.' These are extremely interesting findings that have not been carefully integrated into a philosophical theory of mental time. The findings on the clocks show that the brain represents time by means of mechanisms that work in very much the same way as the clocks humans have designed. And the findings on simultaneity windows reveal that there are at least two ways of conceptualizing the present.

The proposals of this book have surprising and important consequences for philosophy. For instance, there may be forms of temporal representation that lack phenomenal character: they are fundamental for motor control, but are unconscious forms of mental representation. Debates in philosophy that assume that the present is systematically associated with experience may have to be redefined or qualified. Further, a theory of basic justified beliefs for duration may be provided with an assessment of how the clocks work. And finally, attention may be defined as a link that connects the navigational present with the experiential present. Before the analysis of the findings and the argumentation for these proposals, I discuss views about time representation in the first chapter, which canvasses traditional philosophical accounts of time.

CHAPTER ONE

INTRODUCTION

1.1. Minding Time

As mentioned, our awareness of time is a fundamental aspect of our lives and yet, explaining how exactly the human mind is aware of time is a problem that has puzzled philosophers and scientists for centuries. Aristotle asked whether time would exist without a mind that could *count* or *measure* time (Aristotle, 1980, IV, 14). Augustine famously said: "It is in my own mind [...] that I measure time. I must not allow my mind to insist that time is something objective" (Augustine, XI.27). Other philosophers, most notably Kant, claimed that there is an inner *sense* of time (Kant, 2000, 162–163).

These claims capture the intuitive notion that in our daily experience we sense, measure, or count time. But how can one explain these capacities for sensing time? How is it that we represent and experience time: what mechanisms are involved in time representation at the most fundamental level, such that they make possible our immediate acquaintance with time, and how should the representations of time that constitute what we intuitively call 'our immediate awareness of time' be characterized? These are questions that require empirical evidence and philosophical analysis.

One can appreciate the importance of these issues by reviewing the contemporary literature on the philosophy of time, in which similar worries arise. Some views on time hold that while there are properties of time, such as its ordered and asymmetric structure, which are mind-independent, other properties, such as its continuous "passage" or flow, supervene on our cognitive capacities. At the deflationary end of the spectrum of possible views is the claim that no aspects of time are mind-independent and thus that time itself is entirely supervenient on our cognitive capacities.

This extremely deflationary view implies that there is no physical time, and one may think that it is extremely implausible. But actually, some physicists like Julian Barbour (2000) think that time should be removed from physics. Proponents of the complete elimination of time argue that this apparently drastic maneuver is actually compatible with physical theory. Obviously, postulating the inexistence of time poses the significant challenge of explaining *what it is* that we call 'time.' Barbour writes,

> No doubt many people will dismiss the suggestion that time may not exist as nonsense. I am not denying the powerful phenomenon we call time. But is it what it seems to be? After all, the Earth seems to be flat. I believe the true phenomenon is so different that, presented to you as I think it is without any mention of the word 'time', it would not occur to you to call it that. If time is removed from the foundations of physics, we shall not all suddenly feel that the flow of time has ceased. On the contrary, new timeless principles will explain why we *do* feel that time flows (Barbour, 2000, 14).

Indeed, it does seem as if time passes. We feel time passing and we are immediately acquainted with the duration of events. But if time does not exist, what could possibly explain our acquaintance with time? In the philosophical literature J. J. C. Smart, along with many philosophers of time called the 'B-theorists' take a somewhat less extreme view, arguing that only the feature of the flow or passage of time (i.e., that events are first future, then present and then past) is illusory, but that other features of time, such as the concatenation of events by the relations 'earlier,' 'later' or 'at the same time' is real. Even within their view, however, the challenge remains: how to explain our acquaintance with the duration of events that gives rise to the sense of time flowing? Smart is aware of this challenge, and acknowledges that merely pointing out that the passage of time is an illusion is not enough:

> If the passage of time is an illusion it is a strange and intellectually worrying one. It would be good if we could not only give reasons for thinking that it *is* an illusion [...] but also if we could give some sort of explanation of how this illusion arises (Smart, 1980, 10).

I agree with Smart that an explanation of the cognitive basis of our immediate acquaintance with duration and the passage of time is badly needed in the literature on the philosophy of time. Unlike Smart, I will be non-committal as to whether or not the so-called passage of time is a property *of time*. The psychology of time is challenging enough.

Views like those expressed by Barbour and Smart reveal the need to explain the cognitive basis of temporal representation, without making any other assumption about agents or causality. The central role that cognitive capacities for representing time play in reductive accounts of the metaphysics of time (or of features of time) demands a serious effort by philosophers to come to grips with the scientific evidence on time cognition. Philosophers have made tentative empirical claims about time cognition, but very few have taken into account all the current scientific evidence and (as far as I know) none of them has thoroughly reviewed the scientific literature on time representation and the present.[1] Take for instance the following set of claims by Smart:

> We are aware of the flow of information through our short-term memories and we confuse this with a flow of time itself. This conjecture is perhaps supported by empirical evidence according to which the greater the number of stimuli that there are in a given temporal interval the greater is the subjective estimate of the length of that interval. It might be further supported by some more equivocal evidence which suggest that the estimation of the length of an interval depends in the same way on the complexity of the stimuli (Smart, 1980, 13).

Smart does not give details about the evidence he is relying upon. He says that the conjecture he puts forward is *perhaps* supported by the empirical evidence, which he hastily mentions, and it is unclear what he means by 'equivocal evidence.' The connection he makes between the flow of information and the flow of time is uninformative: there are several types of short-term memory, so it is unclear what specific mechanism and set of findings he might be referring to. More importantly, whatever findings Smart has in mind, they are not the *relevant* psychological findings on time cognition (as I explain in chapters 2 and 3, the relevant findings on the representation of the length of intervals concern two cognitive clocks).

Another problem with philosophical accounts of time cognition is that they assume that the sense of time passing is the most critical psychological datum to be explained. However, the 'flow' or 'passage' metaphor is not unambiguous, so it is problematic to pinpoint what the datum is. What exactly is this representation or experience supposed to be and how should it be construed? What is our immediate awareness of time? Are there different forms of representing (and being aware of) time? If so, which are these?

For instance, we seem to be immediately aware of the duration of intervals: we can tell how long intervals are. Our capacity to represent duration and to experience duration in general (these expressions are not

[1] There are some attempts by R. Le Poidevin (2004a, 2007), C. Callender (2008b), and L. Paul (2010) but these are not thorough and sustained investigations of the psychological findings. As mentioned, J. Campbell (1995) refers to research on the circadian clock and related psychological findings, but his main focus is to give an account of the *conscious* self. Thus, the main issue that Campbell focuses on is entirely different from the topic of this book.

synonymous, as the latter necessarily involves phenomenal consciousness) frame our interaction with the world. These cognitive capacities to represent duration are central to explain the so-called "feeling" or experience that time flows. But how do we represent duration? By means of which mechanism(s) are we acquainted with duration? I address these questions in chapters 2 and 3.

There is another aspect of our immediate awareness of time. We seem to be immediately acquainted with the present moment. Bertrand Russell argued that we experience succession (that we perceive successive events) within the *specious present* (an extended moment of time that we *experience* as the present moment). Russell also said that we must be immediately acquainted with very recent past objects through "immediate memory" (Russell, 1913, 70). This also seems to be relevant for the experience of the so-called 'flow of time,' but how should we determine if Russell is right? Are there findings that could explain our acquaintance with the present moment and confirm or disconfirm these claims by Russell (as well as similar claims by other philosophers)? I address these questions in chapter 4.

An explanation of these fundamental and primitive forms of time representation, informed by an assessment of the relevant psychological findings that accounts for how they all fit together in a comprehensive theory of time representation, is something that the philosophical literature is still missing. The two main topics of this book are the representation of duration and the present moment.[2] In the conclusion, I explain the advantages of this book's account of time representation: it provides a scientifically informed and philosophically rigorous model of the representation of duration and the present, which offers original insights for debates in philosophy and psychology.

Two issues concerning the scope of this book deserve special emphasis. First, it will focus on the *representation* of time and not on the nature of time or what time *is*. Some of my proposals on time representation have implications for debates in the metaphysics of time that appeal to our immediate acquaintance with the present, but I will not develop these

[2] Since these are the primitive representations of time that can be clearly defined and for which there is scientific evidence, I shall focus on the representation of the duration of intervals and our acquaintance with the present, and not on the more problematic notion of 'flow.' In the conclusion I suggest that if there were an experience of the flow of time, it would depend on these two primitive representations. Thus, the only way to talk sensibly about the flow of time at the sensory-motor level is in terms of the representations and mechanisms that I assess in the following chapters.

implications because the topic of the nature of time is beyond the scope of the present study. Second, the book will focus on the *sensory-motor representation* of time, i.e., on the most fundamental forms of time representation that ground other mental representations of time, such as those involving judgments of a conscious self, as well as cultural, linguistic or scientific representations of time. In the next section, I spell out in more detail the primitive nature of the sensory-motor representations that I examine in this book, and in section 1.3., I describe its original contributions. Issues concerning the phenomenology of time are discussed in chapter 4.

1.2. Time: The Most Primitive Representation

Of all the experiences and representations that constitute our perception and sensory-motor apprehension of features of the world, our immediate representation of time (a type of temporal representation that grounds other, more complex, representations of time and which does not *depend* on other representations) is unique because it is the most primitive kind of sensorial representation. As Ernst Mach noticed in *The Analysis of Sensations*:

> Much more difficult than the investigation of space-sensation is that of time-sensation. Many sensations make their appearance with, others without, a clear sensation of space. But time-sensation accompanies every other sensation, and can be wholly separated from none (Mach, 1959, 245).[3]

How could one make Mach's point more precise? An area of philosophical research in which the primitiveness of representations is frequently discussed is the philosophy of perception. An essential assumption of any philosophical theory of perception is that there are space-time coordinates that serve as the basis of sensory-motor identifications. This issue comes up vividly in debates about a difficulty concerning the perception of features that are co-instantiated in a perceptual object, called the 'binding problem.'

This problem can be succinctly captured by the question: how does the sensory-motor system bind or put together different features that are

[3] Hans Reichenbach makes a similar point: "The experience of time is allotted a primary position among conscious experiences and is felt as more immediate than the experience of space. There is indeed no experience of space in the direct sense in which we feel the flow of time during our life" (Reichenbach, 1958, 110).

processed independently (such as color and shape in the case of vision) in order to build a unified representation of an object with a specific color and shape? An influential solution to this problem postulates that space-time coordinates serve as the *referents* that the sensory-motor system uses to attribute features (or predicative properties) to objects. Austen Clark (2000) suggests that the capacity of the sensory-motor system to solve the binding problem based on spatiotemporal coordination might be the cognitive origin of the most fundamental linguistic distinction, i.e., the *subject-predicate* distinction:

> If in fact similar collecting principles apply to pairings of place-times and features as to subjects and predicates, then the two phenomena may be related to one another, and the linguistic phenomena of reference and predication may have ancestors in our sensory systems. Nature tends to copy solutions that work, and if aeons ago the ancestors of our visual system (for example) managed to solve the Many Properties problem, it would not be entirely surprising to find that later linguistic systems simply copied their solution. If this were so, then the distinction between reference and predication reflects an even deeper and older architectural feature of the neural organization of our sensory systems (Clark, 2000, 73–74).

I am not endorsing Clark's conjecture about the origin of the linguistic distinction between subject and predicate (although it is plausible), nor am I interested in the binding problem per se. What I want to emphasize is the primitive character of sensory-motor spatiotemporal coordination. Clark nicely captures this idea when he explains how without spatiotemporal coordinates perceptual representation would not get off the ground because no attribution of features to sensorial referents (place-times) would be possible.[4]

Spatiotemporal coordinates are essential for the sensory-motor system to make sense of the features it registers. Although it may not be a sufficient condition for objecthood, spatiotemporal coordination is certainly a necessary condition for the successful individuation of percepts. It is exactly this characteristic of our immediate representations of time (i.e., they are *essential* to interpret sensorial stimuli) that makes them primitive and fundamental. Without these primitive representations of time and their coordination with spatial information, the identification of perceptual objects would be impossible.

Mach's point can now be made more precisely. Space and time representations are fundamental to identify objects in perception, but time

[4] See Clark (2000), particularly section 5.

representation is *even more* fundamental because we can have sensorial representations that are not spatial in character (i.e., they seem to have temporal, but not spatial coordinates). Think of an experience of "pure expectation." It seems that when you are expecting something to happen but you do not know exactly what to expect, your experience may not have spatial coordinates but, as Mach says, it certainly has temporal coordinates i.e., your experience cannot be wholly separated from its temporal representation: how long you have been expecting something.

But if this is so, given that space and time coordinates are the most primitive representations used by the sensory-motor system to refer and attribute features of the environment to objects, and given that time representation is even more fundamental than space representation, then it seems that one must conclude that time representation is the most primitive of all the representations used by the sensory-motor system to interpret *any kind* of information.

Obviously, spatiotemporal representation is crucial to give the sensory-motor system more fine-grained powers of discrimination, and temporal coordination alone would be insufficient for an animal to be a successful interpreter of environmental information. As Clark says, in a 'No-Space' world, different instances of the same quality would only differ temporally and demonstratives would only be used to identify different moments in time. More specifically, information concerning the simultaneous instantiation of qualities at different regions would not be registered, e.g., two identical patches of red would not be distinguished as *two* or in the case of two cat cries "One might count a *series* of cat cries, but not simultaneously sensible cats" (Clark, 2000, 161).

However, temporal coordination alone seems to be (in some cases) enough for successful discrimination (as in the case of only *one* feature that is changing over time). And in many cases, one can think of sensorial experiences that lack spatial information. Actually, one does not have to think of contrived cases in order to come up with a clear example of pure temporal coordination (a case of sensory-motor representation that lacks any spatial character). With respect to olfaction, Clark writes,

> Two simultaneous presentations of an acrid odour fuse to one; and one cannot discriminate the presentation of something that is both acrid and musky from the simultaneous presentations of something acrid and something else that is musky. Of course one can still use one's nose to distinguish an acrid thing and a musky thing from an acrid musky thing, but it requires successive sniffs and a generalization over times (Clark, 2000, 160).

In No-Space olfaction, the sequence of sniffs does not produce a map of locations with sources of odors. It simply produces a series of experiences that have a temporal coordinate, which would allow a creature with the capacity for memorizing sequential sniffs to distinguish and catalogue different odors even in the absence of any spatial information.[5] Thus, time representation is more fundamental than space representation in the sense that there seems to be no *possible* sensorial experience without a temporal character, but there are sensorial experiences without a spatial character (or at least cases in which it is unclear whether or not spatial representation is involved). One can say that No-Time experiences are inconceivable (there is no possible epistemic scenario in which time is not explicitly represented).

I do not mean to suggest that space representation is somehow dispensable in solving the problem of individuating the contents of perception. The notion of a *place-time* is truly fundamental to solve this problem. Place-times provide an ordering of experiences that is unlike any other ordering that could be based on different qualities of sensorial experiences. For instance, as Rudolf Carnap (1967) said, many qualities of sensorial experiences can co-occur at a single place-time sensorial coordinate, but many place-time sensations cannot co-occur in a single sensorial experience.[6]

1.3. Outline of Proposals

Primitive spatial representations can be understood in terms of cognitive maps, but how should temporal primitive representations be construed? In the next two chapters, I explain how there are two clocks (the circadian clock and the stopwatch) that produce metrically structured representations in organisms endowed with a sufficiently complex nervous system. The sensory-motor system uses these representations to produce *spaces of possibilities for action*. An innovative aspect of my account of the clocks is that it provides a general theoretical framework, based on the necessary constraints for any time measurement discussed by Reichenbach (1958), in order to classify interval and periodic clocks, which I then apply to the circadian clock and the stopwatch (chapter 2 is devoted to this issue).

[5] A further assumption for this example to work might be that the creature should never move because in that case olfaction would have a spatial character. But it is not too hard to imagine such a case.

[6] Rudolf Carnap highlights this important difference in (1967, 145–146).

In chapter 3, I explain why the representational outputs of the clocks are metrically structured and propose that these representations are best characterized as analog representations. I postulate five criteria for analog representation and show that the outputs of the clocks satisfy these criteria, relying on experimental evidence to support my arguments. An important contribution of this chapter is that I define the representations of the clocks in terms of their isomorphic mapping to periods and intervals, which specifies non-trivial veridicality conditions for their content.

The explanation of how the two clocks work (provided in chapter 2) is essential to understand how their representational outputs (described in chapter 3) map onto features of periods and intervals. The periodic cycles of the circadian clock allow the sensory-motor system to interpret its outputs in terms of phases of the period of 24 hours. Animals use this information to predict periodic events and navigate. The elapsed intervals emulated by the stopwatch allow the sensory-motor system to interpret moments of an interval in an aperiodic way, thus producing representations of time in a non-periodic fashion. The analysis I offer of these representations provides a reliabilist explanation of the epistemology of time representation (in particular, immediately justified beliefs about duration), which is also an original contribution of this book.

But how are these representations *anchored* to the present moment? I answer this question in chapter 4, which offers a novel two-phase model of the present. There have been a lot of proposals about how the experienced present is specious (not *durationless* or point-like, thus having earlier and later parts), but few have tackled the issue of the specious present in relation to spatiotemporal *coordination*, and there is no account of how the sensory-motor system relates clock representations to the present moment. This is a major gap in current theories of time representation because the third constraint on any time measurement (the specification of simultaneities in relation to the units of time) is not addressed.

The two-phase model of the present offers solutions to these theoretical difficulties. One component of this model, which I call the *sensorial present*, is the relevant anchor for spatiotemporal coordination and the representation of simultaneity. The other component, which I call the *phenomenal present*, unifies conscious experiences. An important aspect of my proposal is that I rely on experimental evidence to justify the adequacy of the two-phase model of the present. I also argue that the distinction between the sensorial present and the phenomenal present may be similar to the distinction between access consciousness and phenomenal consciousness. If I am right about this, current debates about temporal experience need to be thoroughly revised.

CHAPTER TWO

PERIODIC AND INTERVAL CLOCKS: THE UNIFORMITY OF TIME AND THE UNITS OF TIME

The distinction between periodic and interval timing is crucial to approach the issue of temporal representation. It provides a fundamental categorization of representations of time based on the mechanisms that keep track of time. A careful assessment of the differences between periodic and interval timing brings to light aspects of timekeeping that are otherwise confused or blurred, and it should precede any analysis of temporal representation. In this chapter, I provide a general theoretical framework to classify periodic and interval timing based on the essential characteristics of periodic and interval clocks.

The framework I offer in this chapter is of interest to philosophers, psychologists and cognitive scientists in general, for the following reasons. First, it is difficult to have a meaningful discussion on time representation without understanding how timing mechanisms work. Getting familiarized with the specific challenges that measuring time poses is crucial to argue sensibly about temporal representation. Second, I will demonstrate the usefulness of the framework I am offering in this chapter by applying it: I use it to distinguish the main characteristics of the circadian clock and the stopwatch. Although biological clocks have been differentiated through careful experimentation and analysis, there is no unified theoretical treatment of periodic and interval biological clocks. However, many of the ideas I employ in developing this theoretical framework, and a significant portion of the evidence I rely upon, come from C. R. Gallistel's work, particularly Gallistel (1990).

Third, I seek to clarify in this chapter how periodic and interval clocks, which are the basis for temporal representation in humans and animals, satisfy the basic constraints that any measurement of time with specific accuracy or veridicality conditions must satisfy. This is extremely important because I will not rely on other concepts, such as causation or the self to account for primitive forms of time representation (as mentioned, this is an advantage over other accounts such as Campbell's). Rather, I will show how by satisfying these constraints, such representations have content that is independent of conceptual repertoires and nonetheless provides an epistemic basis for basic belief-forming processes in humans, and gives immediate information of the physical environment to animals.

Thus, the significance of these constraints is that such representations map directly onto the environment without the need of a complex web of inferential relations among concepts. Actually, part of how these constraints *guarantee* that these representations have accuracy conditions is by isolating the content of these representations from conceptual representations (for instance, of a theoretical sort). Reichenbach (1958) is quite explicit about the importance of what he calls 'coordinative definitions.' The essential characteristic of coordinative definitions is that they are not defined in terms of other concepts. Rather, they specify a unit of measurement by referring directly to the physical world. The same holds for mental representation (i.e., some representations are defined in terms of their relations to other representations, while others, such as the temporal representations I shall focus on, are defined in terms of what they refer to in the environment).

Fourth, there are original contributions of my analysis of periodic and interval clocks, for instance concerning how they relate to each other and what principles govern their interaction. And fifth, with respect to methodology, I follow as closely as possible the scientific evidence. When I describe the most general features of periodic and interval clocks, I avoid speculation by exemplifying these features with real clocks that were designed throughout history.

One of my goals is to demonstrate the existence of two types of sensory-motor representations of time in animals and humans, based on the framework I offer to distinguish between periodic and interval timing. I address the topic of sensory-motor representations of time in the next two chapters. In this chapter, I present the theoretical framework for distinguishing periodic and interval timing, and demonstrate that the periodic and non-periodic processes emulated by the circadian clock and the stopwatch satisfy the requirement of the uniformity of time. This solves the problem of specifying the *congruence* of time periods and intervals, and also the requirement concerning the definition of *units of time* that can be used for measurements of time (i.e., the phases of cycles or the segments of intervals).

The structure of this chapter is as follows. Section 2.1. provides a general description of different techniques for measuring and calculating time, which motivates the distinction between periodic and interval clocks. Section 2.2. provides a theoretical account of the main characteristics of periodic clocks and section 2.3. offers a theoretical account of the characteristics of interval clocks. In section 2.4., I compare periodic and interval clocks, emphasizing their advantages and disadvantages for time measurement. In section 2.5., I apply my characterization of periodic clocks to describe the circadian clock and section 2.6. is a description of the stopwatch, based on my characterization of interval clocks.

2.1. Records of Time

A sundial projects a shadow onto a disc or some type of base. This shadow keeps track of the trajectory of the sun from sunrise to sunset. The object that casts the shadow, also known as *Gnomon*, needs to be tall enough to cast a shadow that reaches the end of the circumference of the disc (or the relevant portion of the base) and it has to be at a particular angle (which is the latitude of the location in vertical sun dials) with respect to the disc or the base in order to cast a shadow continuously. The *Gnomon* also has to be thin enough to cast a narrow shadow that can be projected along the hour lines, which are numbered lines drawn on the disc or base.

Sundials are very interesting pieces of technology. They reveal mankind's need to control and understand time, as well as the enormous influence that representations of time have on our daily activities. It was because of the development of precise sundials that we divided the day into hours. The period of 24 hours is a natural cycle—it marks the period of one complete rotation of the earth on its own axis, or more colloquially, the period that elapses from dawn to dawn. From a completely geocentric (and erroneous) perspective, 24 hours is the duration of the cycle in which the sun "travels" or makes its way through the sky to start a new cycle of daylight.

The period of 24 hours is one of the most important temporal cycles for many creatures living in our planet, and some have considered it to be the most practical natural clock for time measurements in general (See Reichenbach, 1958, 121). Below, I describe the circadian clock, a biological mechanism that keeps track of time periodically, in cycles that approximate 24 hours. The period of one day is so crucial in our lives that we divide it carefully into units (e.g., hours, minutes and seconds). However, there are other important temporal cycles that are registered in many ways, each of them as fascinating and revealing as the way in which a sundial records time by projecting a shadow that emulates the trajectory of the sun across the sky, or the way in which the circadian clock mimics the period of one day by means of cyclical biological processes. A thorough assessment of natural clocks must account for all kinds of periods and processes.

Tree rings record the passage of time periodically, in cycles of about one year. These rings can be seen in any tree if one cuts across its trunk horizontally. They form because of periodic changes that depend on the seasons, which have an impact on the growth of the tree, e.g., temperature and rain. As the tree grows and creates new wood, the periodic changes of the seasons are recorded in each of its rings and, thus, each ring demarcates a period of one year.[1] A very similar process occurs in ice, where seasonal changes in temperature and sun irradiance are registered in the layers of ice cores, as ice accumulates. These layers, like the rings of a tree, correspond to a period of one year.

Some registers of time cover much wider temporal ranges. Carbon-14 decays into Nitrogen-14 very slowly, within a scale of thousands of years. The rate of decay is quite reliable (Carbon-14 has a half-life of 5730 years) and it allows scientists to calculate the time at which a now fossilized creature was a living and thriving organism. But light (radiation in general) is the most amazing record of time with respect to temporal range. The speed of light is constant and if one calculates the distance that light has traveled from its source, one can calculate the time it has traveled, because speed is distance divided by time. Cosmologists can calculate the distance that light has traveled because of the redshift of light (a shift in the wavelengths of light from higher to lower energy) produced by the accelerated expansion of the universe.[2] Using the redshift of the spectrum of light, scientists can estimate the time that light has traveled from extremely distant sources, making possible calculations of time in the range of millions and billions of years.[3]

---

[1] The technique for dating based on the study of tree rings is called *dendrochronology*. Dating based on tree rings has been a powerful tool in determining the atmospheric conditions of ancient environments.

[2] There are three kinds of redshift: Doppler, gravitational and cosmological. All are relevant for calculating time, but the most relevant for the present purposes is the cosmological redshift, which is caused by the expansion of the universe. There is an opposite process to redshift, which is particularly important to the Doppler effect. If the wavelengths are shortened, from lower to higher energy (which means the distance between source and observer is decreasing instead of increasing) the shift is called blueshift.

[3] The last calculation of the age of our universe indicates that it is approximately 13.73 billion years old. This estimate is based on the WMAP (Wilkinson Microwave Anisotropy Probe), and it is considered to be the most precise measurement of the age of the universe to this day. Cosmologists use the redshift technique to map the universe, identify some of the most ancient cosmic events and the most distant galaxies.

These dramatically different scales of time (i.e., days, years, thousands of years, millions of years) are a source of wonder. But equally astonishing is the fact that time is recorded in nature in so many different and *reliable* ways. One can think of all these natural processes—the cycle of day and night, the seasons of a year recorded in tree rings and ice cores and the processes that lead to carbon dating and the redshift of light—as *natural clocks*. But there is a noticeable difference between the *cycles* of one day and one year, registered by the circadian clock of living organisms and by tree rings and ice cores respectively, and the long *intervals* of time registered by Carbon-14 and light.

Clearly, one could add the number of tree rings of a specific tree and calculate an interval of time in terms of years. But the process on which the registration of time depends upon is cyclical—it repeats itself over and over. This is manifestly not the case with Carbon-14 and light. There is no cycle on which the decay of Carbon 14 or the redshift of light depend. These different forms of timekeeping are very important to understand two fundamental ways in which animals and humans represent time. The remainder of this chapter is devoted to clarifying the distinction between periodic and interval clocks and to expanding on the characteristics of periodic and interval timing.

2.2. Periodic Clocks

In this section, I explain the most important characteristics of periodic clocks and some of the advantages they have as registers of time. In section 2.3., I describe the characteristics and advantages of interval clocks and in section 2.4., I offer a comparison between periodic and interval clocks. This provides an important theoretical background that is necessary to understand how the circadian clock and the stopwatch satisfy the constraints concerning congruence and units of time required to accurately measure time. Periodic and interval clocks perform the same function—they register time. But they perform this function in very different ways. In order to appreciate the importance of the distinction between periodic and interval timing, one must first understand the nature of the mechanisms that underlie these forms of timekeeping. As C. R. Gallistel says,

> Mechanisms or processes that make possible the recording of moments in time and the determination of temporal intervals come in two basic forms: oscillatory processes and nonoscillatory decay or accumulation processes (Gallistel, 1990, 231).

Oscillatory processes and the mechanisms that produce them underlie periodic timing. These processes repeat in a regular fashion and each repetition of a cycle occurs in accordance with a *constant* period of time that delineates the beginning and the end of the cycle: the end of one cycle marks the beginning of a new cycle. The repetition of constant periods satisfies the *uniformity of time* constraint (each period has *congruent* phases and durations). According to this definition, the rotation of the earth on its own axis, the rotation of the moon around the earth and the rotation of the earth around the sun are oscillatory processes. These processes repeat with a regular period and the end of one cycle marks the beginning of a new one.

These periods (a day, a lunar month and a year) are quite familiar to us.[4] However, other oscillatory processes have much shorter or longer cycles. Examples of these cycles are the oscillations of a cesium atom (the mechanism underlying atomic clocks) and the rotation of the sun around our galaxy, or cosmic year. The cesium atom produces around 9 billion oscillations per second and the sun makes one rotation around the galaxy about every 200 million years. In spite of the significant differences in scale and physical instantiation, all these oscillatory processes share the following characteristics that make them *reliable clocks*:

a) There is a period of time that delineates each cycle of the oscillatory process.
b) The cycles repeat themselves one after the other in accordance to a constant or uniform period of time.[5]
c) There is no restriction concerning the physical medium that instantiates an oscillatory process. What characterizes an oscillatory process are properties a) and b).[6]

Properties a) and b) are the foundation for the most important characteristic of periodic timing mechanisms, i.e., that the oscillatory mechanism is always at a particular *phase*. The phase of an oscillatory process is the stage of completion of one cycle. Determining the phase of an oscillatory

---

[4] The period of the lunar month varies depending on how it is measured. Roughly, it is between 27 and 29 days.

[5] Notice that the period of a cycle is very different from the frequency of an oscillatory mechanism. The frequency of an oscillatory system is a measurement of the number of cycles per unit of time. This unit of time is arbitrarily selected (it could be one second or one year) and it can differ substantially from the period of time in which the oscillatory mechanism completes one cycle.

[6] I explain this property below, particularly in section 2.4.1.

process is equivalent to establishing how much of the cycle has been completed. If one characterized a cycle as a circle, then the end of the cycle would correlate with 360 degrees (which is also the beginning, or 0 degrees), and the intermediate phases of the cycle could be characterized by the intermediate degrees that range from 1 to 359. The phases of a uniform cycle satisfy the main constraint imposed on units of time, because they can be used to measure or count time. The uniformity of the cycle guarantees that these units have the same duration.

The fact that oscillatory processes and mechanisms are always at a particular phase is crucial to understand why such processes and mechanisms register time in a reliable way.[7] For instance, if two or more oscillatory mechanisms with the same periodic cycle are *phase-locked* (which means that these mechanisms share the same phase) then the phase of one such mechanism reveals the state of the whole system of phase-locked oscillatory mechanisms.[8] Moreover, oscillatory mechanisms can have different periodic cycles and yet maintain a fixed relationship. Gallistel explains why the hands of a clock are a good example of a fixed-phase relationship among cycles as follows:

> On a clock, the period of the second hand is 1 minute, the period of the minute hand 1 hour, and the period of the hour hand 12 hours, but the phase relationship among these cycles is fixed. The second hand completes its circle just as the minute hand indicates the minute, and the minute hand completes its circle just as the hour hand indicates the hour (Gallistel, 1990, 229).

Thus, the phase of an oscillatory mechanism is important to determine synchrony among oscillators with equal periods and also to fix relationships among oscillatory mechanisms with different periods, which is illustrated by the way in which the cycles of the hands of an analog clock are related. The phase of an oscillatory process makes possible the precise

[7] In what follows, I will be mostly concerned with discussing oscillatory *mechanisms* because of their relevance for understanding the circadian clock. Processes are more pertinent to the discussion of interval clocks. I do not mean to draw a sharp distinction between processes and the mechanisms that underlie them. I am distinguishing one from the other based on the intuitive notion that mechanisms always produce the same result (just like oscillators always produce the same cycle). 'Process' is a more general category, not inherently related to reproducibility, which I think suits better what I have to say about interval clocks. In any case, nothing that I have to say hangs on this distinction.

[8] If two oscillatory mechanisms are *coupled* (which means that the phase of one mechanism changes the phase of the other, and their phase is different) then what determines their phase will be the dynamics between these mechanisms. This means that the relationship between their phases is not fixed and it changes with time.

coordination of its cycle with the cycles of other oscillatory processes. This is why such coordination is the most important characteristic of oscillatory processes with respect to registering time. By phase-locking their cycles and fixing phase relationships, oscillatory processes can reliably register time in many scales and in many different ways. By reproducing a cycle with a constant period, oscillatory processes are clocks that keep track of time periodically. To conclude and reiterate, the most important properties of periodic clocks (which are the defining properties of oscillatory processes), are:

a) There is a period of time that delineates each cycle of the clock.
b) The cycles of the clock repeat themselves one after the other in accordance with a constant or uniform period of time.
c) There is no restriction concerning the physical medium that instantiates a periodic clock.
d) The clock is always at a particular phase.

These properties, all of which must be satisfied by a mechanism or process in order to qualify as a periodic clock, differ significantly from the properties that characterize interval clocks. The importance of the distinction between periodic and interval clocks will be evident in the discussion of the psychological data regarding timing mechanisms in animals and humans. However, before assessing the psychological evidence, it is important to distinguish periodic and interval clocks from a purely theoretical perspective. Having at hand the characteristics of periodic clocks makes much easier the task of defining interval clocks, which is the main goal of the next section.

2.3. Interval Clocks

Unlike periodic clocks, interval clocks are best described as "one-time" processes because there are no repetitions of cycles or phases. This crude characterization can be made much more precise by comparing the main characteristics of interval clocks with the properties of periodic clocks. But first, before listing the characteristics of interval clocks, a couple of illustrations of the type of process that instantiates them would be useful. I will focus on two classic examples of interval clocks: the sand clock (or hourglass) and carbon dating.

The process by means of which a sand clock registers time is known in the literature as a non-oscillatory *accumulation* process. It is non-oscillatory because it does not depend on the periodic repetition of cycles in order to register time. The stages of time-registration of a sand clock are not *phases* of a cycle, but rather *partitions* of amounts of sand that accumulate in the lower part of the sand clock.[9] The interval of time registered by the sand clock depends fundamentally on its spatial capacity, i.e., it depends on how much sand it contains, or how big it is. Sand clocks depend on an accumulation process because the amount of time registered correlates with the amount of sand that builds up at the bottom of the clock. Sand clocks need to be "started" by turning them around. The sand starts falling through the narrow tube of the clock and accumulates in its lower part, indicating how much time has elapsed. Generally, sand clocks have no partitions or lines that could indicate increasing intervals of time, although it is in principle easy to construct such a sand clock. However, most sand clocks can only register whatever time interval can be correlated with the total amount of sand that the clock contains.

There are important physical constraints that sand clocks must satisfy in order to be reliable. For instance, the sand must flow *continuously* in order to satisfy the uniformity of time constraint. This means that the sand must be thin (in proportion to the size of the clock). Otherwise there is the risk that the sand could obstruct the tube, which would slow down the flow of sand, altering the measurement of the interval of time. Other physical aspects that change the constant flow of sand are the angle of the glass bulbs or containers, the width of the tube (which must also be uniform), the size of the sand particles, and the position of the clock (it must be on a flat surface, otherwise it slows down).[10]

Thus, what makes a sand clock reliable is its *constant rate of sand-flow*. If this rate of flow fluctuates or changes, for instance because one of the previously mentioned requirements was not satisfied, then the clock is unreliable: it will speed up and slow down randomly and measure different intervals each time it speeds up or slows down, thereby making it impossible to satisfy the uniformity of time constraint. Once one considers all these physical constraints on the design of a sand clock and, more importantly, how easy it is for a sand clock to speed up or slow down, it is not surprising that we only use them today to boil eggs or play board

---

[9] I have in mind the typical sand clock with a symmetrical shape, in which sand falls from the top part of the clock to the lower part through a narrow tube.

[10] I am leaving out other equally important, but less obvious factors that affect the accuracy of sand clocks by altering their constant flow of sand. These factors are temperature, moisture, the density and weight of the sand particles and the smoothness of the surface of the glass.

games. But this does not mean that all interval clocks share these problems.

Other accumulation processes, like the formation of layers of rock that serve as the basis for calculating the age of the earth, have a relatively steady rate of sedimentation. There are, of course, climatic factors that change this rate.[11] But, in general, the rate is much less dependent on specific constraints, like those that determine the flow of sand in a sand clock. Another sedimentation process with an even more reliable constant rate is the Erythrocyte Sedimentation Rate (ESR). It determines the rate at which red blood cells (those that carry oxygen) accumulate at the bottom of a test tube containing blood. The reliable rate at which red blood cells fall to the bottom of the test tube allows doctors to diagnose blood-related conditions, particularly inflammation. The diagnosis is based on whether or not the rate is the standard one.

Interval clocks can also be instantiated by *decay* processes. As mentioned, Carbon 14 is a very reliable interval clock because its rate of decay, though extremely slow, is quite consistent. The non-oscillatory decay process that serves as the basis for carbon dating does not have many of the problems that jeopardize the accuracy of sand clocks. First, there are no constraints with respect to position and other environmental situations that severely affect the rate of flow in the sand clock. If "left alone," Carbon 14 naturally decays into Nitrogen 14, and it is incredible how precise its rate of decay is—very few environmental conditions affect its rate of decay.

Decay processes are as diverse as accumulation processes. There is radioactive decay in many particles, besides Carbon 14, and many fermentation processes occur at a specific rate. A metaphorical, though not very precise, way of describing the difference between accumulation and decay processes is as follows. Accumulation processes are evidence that time always goes forward, making the past inexorably bigger and bigger. Decay processes are evidence of the inexorable passage of time too, but they show how time destroys everything it touches.[12] They are two faces of the same coin, but one emphasizes how the past grows and the other emphasizes how the present is ephemeral.

[11] I clarify, in section 2.4.3. of this chapter, why tree rings and ice cores are related to periodic clocks, and also why the relatively constant rate of accumulation of layers of rock makes rock formations *interval* clocks.

[12] Candle clocks, which measure intervals of time in terms of the rate at which a candle burns away, are a very good illustration of a decay process in which time destroys the medium used to measure it.

Accumulation and decay processes are the basis of all interval timing. I have mentioned some of the characteristics of interval timing. I shall now make explicit these characteristics and contrast them with the properties of periodic clocks. The main characteristics of interval clocks are:

a) Interval clocks have a uniform rate of accumulation or decay.
b) There is an activation event that starts the clock (an event that starts an accumulation or decay process). An interval is measured either by stopping the clock or by using its rate to deduce an interval.
c) There are restrictions on the physical medium that instantiates an interval clock. The most important one is that its scale or size is inversely correlated with its resolution.
d) Interval clocks register *segments* of an interval, not *phases* of a cycle.

I proceed to explain these properties, all of which must be satisfied by a physical medium to qualify as an interval clock. Property a), the constant rate of interval clocks, is what makes them reliable timekeepers and what satisfies the uniformity of time constraint. I will illustrate this point with a hypothetical sand clock. Suppose that you live in ancient Rome—where sand clocks were apparently used to time speeches—and you are commissioned to design a very precise sand clock, so that all the very important Roman politicians get exactly the same amount of time to give their speeches (i.e., intervals measured by this clock will be *congruent* and have the same duration). You know a lot about sand clocks, and proceed to get the best materials—the right kind of sand, glass, wood for the bases, etc. Once you are done, how do you test if the clock is reliable? You could start using the clock to time many things, like concerts, theater presentations, and so on. You could also compare it to other very reliable clocks for measuring the same interval and see if they match by finishing at the same time. But this will not suffice to guarantee that the clock is always measuring the exact same interval. For it is possible that the concerts and theater presentations have quite different durations, just like other sand clocks might be measuring quite different intervals. Your clock might be matching these different durations and intervals because its rate of sand-flow is fluctuating. So, the problem is how to control for fluctuations in the rate at which the sand flows?

The only solution is to tackle the problem directly—the clock has to be stopped at different moments and the amounts of sand that are falling at different moments need to be measured. Before sealing the clock by attaching the bases to the glass structure, you need to determine how much sand is flowing, say at three different moments: when the clock is full, half full and almost empty. You go ahead and in very brief amounts of time marked by a drumbeat played by a musician (for the sake of simplicity, assume that these brief intervals are one second long) you collect the amount of sand that fell from the top of the clock. You verify, to your satisfaction, that at each moment, marked by the drumbeat, you collect 50 particles of sand. This means that the rate of sand-flow is constant, i.e., 50 particles per second.[13] Now you know that no politician is going to worry because the clock is flowing faster when it is full or is changing its rate randomly. This example illustrates that an interval clock's accuracy depends entirely on its constant rate of accumulation or decay.

Property b) can be explained in more simple terms. With respect to information, all interval clocks give the same kind of output: they calculate an interval from a starting time $t1$ to another time $t2$. These moments in time coincide with the activation and deactivation of the interval clock. In the case of a sand clock, $t1$ would be the moment in which the clock is turned around, so that sand can start falling, and $t2$ would be the moment in which there is no more sand at the top of the clock. Regardless of how long or short the interval is, all interval clocks are activated and deactivated, and these events coincide with the beginning and the end of the interval measured by the clock.

The property that makes a manifest physical difference between periodic and interval clocks is property c). There are two aspects of property c), which I will label C1 and C2—and both of them are very relevant to the physical instantiation of interval clocks. C1 concerns a general constraint on the *resolution* of interval clocks and it can be stated as follows: there is an inverse correlation between the scale of the durations that an interval clock can measure and its resolution—the larger the scale of the intervals, the poorer the resolution of the clock, and vice versa.

Think again about the hypothetical sand clock. You have an extremely accurate sand clock that has a very reliable constant rate. But now you want to determine how long people are clapping to one of the speeches. You could use the speech-clock and stop it when people stop clapping. Then you can estimate the proportion between the interval that corresponds to the amount of sand collected at the bottom and the interval that the clock calculates. You can actually count the sand particles in order to be more precise—you know that the rate is constant, 50 particles

[13] Small deviations from this rate, such as 53 particles per second, will not significantly affect the accuracy of the clock. Obviously, how much deviation is allowed depends on how accurate one wants the clock to be.

per second. Counting the particles would give you a very good approximation to the exact duration of the interval that elapsed.

But clearly, this is very messy. Counting the particles means that you have to disassemble the clock every time you want to measure a smaller interval. Stopping the clock takes some time, which will be unfortunately added to the interval that you are trying to measure. You may try to compensate this by anticipating the end of the clapping, but any anticipation will subtract time from the interval being measured. Moreover, the clock has to be tilted every time you stop it, which means that the flow of sand slows down and its rate changes. Thus, your approximation will never be as good as the precise measurement of the interval that the clock was *designed* to calculate.

Things get much worse if the intervals to be measured are very small. The smaller they get, the worse your approximations will be. Actually, if you wanted to be ambitious and calculate intervals of 10 milliseconds, you would find out that sand particles are too coarse to calculate such intervals. So, in spite of its accuracy, the sand clock is useless to calculate very short intervals. Now suppose that you build a "dust" clock for very short intervals. Its resolution is great: the particles are extremely small, which allows you to measure ever-smaller intervals of time. But now how do you go about measuring the duration of speeches, or banquets with the short-scale clock? The scale/resolution tradeoff, property C1, is an insurmountable problem for all interval clocks. It is also instantiated in decay processes. Carbon 14 is reliable for very long intervals of time, but not for speeches or banquets.

C2 is deeply related to C1. Actually, C1 is a consequence of C2, which is a constraint on the type of process that underlies interval clocks. Accumulation and decay processes are ubiquitous in nature. Those that count as interval clocks must comply with property a), i.e., they must accumulate or decay at a constant rate because otherwise they cannot comply with the uniformity of time constraint for the congruence of intervals that all clocks must satisfy. This rate depends on increases or transformations of a *medium*, which could be any physical substance or material. In the case of accumulation processes, there is an increase in the amount of a particular substance or material. And in the case of decay processes, there is a transformation of a substance or material into something else. C2 is a very significant constraint on interval clocks: the medium that instantiates the clock must be such that it either accumulates or decays at a constant rate, and this substantially reduces the set of possible *materials* that can serve as interval clocks.

The constraints on interval clocks explain why periodic clocks are more abundant in nature and less prone to fluctuations in their accuracy caused by changes in the media that instantiate them. Take, for instance, the period of one day caused by the rotation of the earth on its own axis. There might be many transformations in the atmosphere and material constitution of the earth, but the period of its rotation will not fluctuate because of these changes. The medium of a pendulum, which experiences *no accumulation or decay processes*, is a paradigmatic example of a periodic clock. Thus, unlike an interval clock, there is no restriction concerning the physical medium that instantiates a periodic clock, which is property c) of periodic clocks.

The final property of interval clocks, property d), satisfies the constraint regarding the *units of time*. The activation process that marks the first moment of the interval being measured, or t_1, is also the beginning of a *measurement* that will conclude at some point in the future. The two extreme moments of this measurement, t_1 and t_2, delineate the moments of activation and deactivation of the clock. There is no periodicity or repetition of any kind involved in this measurement. It is simply an expanse of time that is best described as a *line*, rather than a *circle*. Since no cycle is involved, partitions of this measurement are segmentations of an expanse of time, not phases of a cycle, which should be equal in length in order to be used as the units of time. This is going to be very relevant for issues concerning the representation of time based on periodic and interval clocks, which I address in the next chapter.

It should be clear at this point that periodic and interval clocks operate in very different ways. Humans have used both of them to measure time. Sections 2.5. and 2.6. of this chapter are devoted to the fascinating scientific evidence that has identified two clocks that animals and humans routinely use to measure and represent time, which can be categorized as periodic and interval clocks respectively.

Periodic and interval clocks correlate with the cyclical and linear conceptions of time, which have generated so much debate in historical, sociological, and anthropological studies. The distinction between cyclical and linear conceptions of time is also of interest to scientists, because the asymmetry of time that generates a direction (or arrow of time) from the past towards the future only makes sense in linear time. If time loops back in a cycle and time travel is possible, then no distinction between past and future is tenable.

I shall now proceed to compare periodic and interval clocks and then examine the psychological data on the circadian clock and the stopwatch in the light of the distinction between periodic and interval clocks.

2.4. A Comparison Between Periodic and Interval Clocks

In this section, I discuss some of the advantages and disadvantages of periodic and interval clocks, focusing, for the sake of simplicity, on the disadvantages. It is important to compare periodic and interval clocks for two reasons. The first is that it clarifies some of the fundamental differences between periodic and interval clocks, and the second is that comparing these types of clocks is extremely useful to understand the representational properties of the circadian clock and the stopwatch.

I first describe, in subsections 2.4.1. and 2.4.2. the disadvantages of periodic clocks and interval clocks respectively. In subsection 2.4.3., I address the issue of *interfacing* periodic and interval clocks. I introduce two principles governing the interaction between periodic and interval clocks that will be very important for sections 2.5. and 2.6., where I discuss the characteristics of the circadian clock and the stopwatch.

2.4.1. *Disadvantages of Periodic Clocks*

There are three disadvantages of periodic clocks, which concern mainly property c) (unrestricted media), but relate to all other properties. The first disadvantage is that periodic clocks do not capture, as well as interval clocks, our intuitive notion of time. This is a disadvantage because the *visualization* of certain properties of time that philosophers and scientists use or find intuitive, like the linearity or the passage of time, is not facilitated by cycles and their representations. Second, periodic clocks do not convey information about the magnitude of durations explicitly. And third, these clocks confound different moments in time, which is a well-known shortcoming of periodic timing. I proceed to explain these disadvantages.

Interval clocks are analogues of the temporal durations that they measure.[14] Simply put, as the interval gets larger, so does the medium that instantiates an accumulation-interval clock. And in the case of decay-interval clocks, as the interval gets larger, the closer one gets to the transformation of the medium into something else. An example of this analogue way of registering time in the case of an accumulation clock is the sand clock, which I have discussed in some detail. In that case, the analogues of durations are amounts of sand: larger durations correspond to larger amounts of sand.

[14] I am using the term 'analogue' in a non-technical way. In the next chapter, I will define *analog* representations, and I will use the term 'analog' in a technical way.

A decay clock that illustrates nicely this type of analogue interval registering is the candle clock. Instead of increasing proportionally to an interval, its medium (wax) decreases proportionally. By melting away and transforming its shape, the thin candle that instantiates the clock exemplifies quite literally how time passes, moving forward relentlessly. Numbered marks on the candle delineate discrete intervals, or units, of time. As the candle melts away, these marked portions of the candle progressively disappear, which allows one to stop candle clocks at a certain section of the interval that they measure.

Periodic clocks are also analogues of the cycles that they emulate. But periodic clocks are always repeating the same phases within a cycle. The media of periodic clocks do not get bigger as durations elapse, or decay accordingly. This aspect of interval clocks, i.e., the fact that their media grow or decay proportionally to an interval, facilitates visualization, and it appeals to our intuitive understanding of time. For instance, it illustrates the intuitive ideas that time flows at a particular *rate*, that it advances towards the future, that the past gets constantly bigger, and that the present moment is ephemeral. Periodic clocks have the disadvantage that they do not capture as well as interval clocks these intuitive notions.

The second disadvantage of periodic clocks is very related to the one I just mentioned. The visualization of intuitive properties, which we attribute to time and which interval clocks make possible, includes the visualization of the magnitude of intervals. This disadvantage can be stated in terms of information registration as follows: periodic clocks are not adequate to convey information about magnitude *explicitly*. This is a significant problem because information about magnitude is fundamental to define metric properties of time. For instance, it is common to define time as one-dimensional and to compare its structure with the metric structure of a line that can be segmented into ever-smaller portions, just like an interval of time can be segmented into ever-smaller intervals.

The similarity between the structures of a line and an interval is manifest, or explicitly encoded, in the media of interval clocks because a line can be considered as an idealization or geometric representation of the physical process of accumulation or decay of an interval clock. The distance that exists between different moments of an interval is explicitly encoded in the *medium* that instantiates the clock, as is illustrated by candle clocks. The explicitness of this encoding of information about the magnitude of an interval, which periodic clocks lack, is an important representational advantage. Another way of saying this is as follows: knowing how much time has elapsed requires information about the beginning of an interval. This information is explicitly encoded in interval clocks because the first unit of time of the interval being measured corresponds to the moment of the activation of the clock. Information about the beginning of an interval is *not* encoded in periodic clocks. Knowing that a periodic clock is at a particular phase tells you nothing about the beginning or end of an interval.

However, there is a way in which periodic clocks can be *used* to calculate intervals (after all, periodic and interval clocks are both reliable time-keepers). No phase of a periodic clock *per se* gives information about when an interval started, but if one obtains this information through other means and knows how it relates to the cycles of a periodic clock, then it is possible to calculate intervals (for example by counting the phases or units of time and the cycles of the clock). But it is a real disadvantage of periodic clocks that information about *when* an interval started is not registered in any of the phases of its cycle.

Finally, periodic clocks are not adequate to individuate specific moments of time by distinguishing them from other moments. The recurring phases of cycles blur distinctions about specific moments that correspond to such phases. This disadvantage is similar to the previous one, except that the limitation consists not in the lack of information about the *magnitude of an interval*, but in the lack of information about *individual moments* in time.

No phase of a periodic clock *per se* gives information about individual moments, because any particular phase of the cycle corresponds to an unknown and, in principle, infinite number of individual moments. If one selects a phase of a particular cycle one can effectively convey information about a moment in time, but one cannot label or identify such a moment based on that phase of the cycle. The problem is that the *same phase* picks out many other moments that correspond to other cycles of the periodic clock. So, knowing that the cycle is at a particular phase gives no information about any *specific* moment, as the following example shows.

You can use the phase 'sunrise' (a phase of the periodic clock instantiated by the rotation of the earth on its own axis) to convey information about a moment in time, such as the moment you were born. Knowing that it was sunrise when you were born does not tell you anything about *how long ago* you were born or *when* your birthday is. Actually, knowing that it was sunrise when you were born gives you as much information about the moment of your birth as knowing that it was sunrise when a Neanderthal woke up informs you about the time in which Neanderthals were alive. Phases do not convey such information and, for all you know, these two events (your birth and the Neanderthal's awakening) might have happened at the same time. However, if one has access to information about specific moments through other means—i.e., by using a memory register, or a calendar—then one can *use* a periodic clock to identify individual moments in time.

By contrast, interval clocks distinguish moments explicitly by accumulating or decaying at a specific rate. Going back to the previous example, if one had an interval clock of a large enough scale, like Carbon 14, one could differentiate sunrise1, when the Neanderthal woke up, and sunrise 2, the moment of your birth, as two distinct moments in time. Then one could differentiate the time in which Neanderthals were still alive and our times.

To summarize, periodic clocks have three important disadvantages as time registers:

1. They do not capture our intuitive understanding of time.
2. They do not convey information about the magnitude of durations explicitly.
3. They confound different moments in time.

2.4.2. *Disadvantages of Interval Clocks*

There are four limitations of interval clocks. The first is that there is always a scale/resolution trade-off in interval clocks. The second is that they cannot be related to one another in a *reliable mechanical* way. This imposes significant limitations on the possible ways in which a set of interval clocks can be manipulated. Third, interval clocks lack sensitivity to periodicity. And fourth, they lack the kind of automaticity that could mechanize behavior, which makes them *user dependent* timekeepers. I proceed to explain these disadvantages.

As mentioned, a very significant disadvantage of interval clocks is that there is always a scale/resolution trade-off. I described in section 2.3. how this problem affects interval clocks, using the example of two sand clocks with different scales (one for debates and the other for very brief intervals). Periodic clocks do not have this problem. As long as their period remains constant, they are incredibly reliable and, since they repeat their cycle regularly, it is possible for them to cover very large, as well as extremely small, scales. Since their media do not have to emulate an interval, scale is not a limitation. Periodic clocks with very short periods can cover vast amounts of time because their cycles repeat for vast amounts of time.

Another crucial disadvantage of interval clocks is that they cannot be related to one another in a reliable mechanical way. As I discussed in section 2.2., there are at least two fundamental ways in which different periodic clocks can be related. The first is by being *phase-locked*: if many periodic clocks are in the same phase and their cycle has the same duration, then the state of the phase of one clock reveals the state of the whole system of phase-locked clocks. Discovering that a set of clocks are not phase-locked may also be important and informative, because then one can determine the lack of synchrony within a system, which may reveal interesting dynamics between the oscillatory mechanisms of the periodic clocks. The other way in which different periodic clocks can relate to each other, even if they have different periodic cycles, is by maintaining a fixed relation, which I also described in section 2.2. In the case of interval clocks, there is no genuine relationship that could resemble the relations between the phases and cycles of periodic clocks. Even if one tried to build a contrived multiply-layered system of interval clocks, the end result would be a single contrived *interval clock* that complies with all the characteristics of interval clocks I mentioned in section 2.3. This is in sharp contrast with periodic clocks. One does not end up with a contrived periodic clock when one has a system of phased-locked clocks or a set of clocks with a fixed relationship: each clock keeps its own cycle and the relations among them are genuine ones. I proceed to explain these claims further.

For the sake of thoroughness, I will give two illustrations of contrived relations among interval clocks in order to show how these relations are not genuine, in the sense that they are not reliable and mechanically reproducible. Imagine that you want to build a system of synchronous interval clocks, the equivalent of a phase-locked set of periodic clocks. All the interval clocks are reliable and flow at a constant rate. Their media could include sand, wax and water.[15] But, in order to simplify things, suppose that they are all sand clocks made with the same materials. How are you going to achieve synchrony among these clocks?

Because of property b) of interval clocks, they have to be activated at the same time. But then the synchrony is not achieved by the clocks, but rather by *your activating them* at the same time. What if you want to keep in synchrony 1000 interval clocks? That would be a nightmare. In contrast, two, or a hundred or a million periodic clocks might just

[15] Water clocks are as old as sundials, probably one of the first ways in which humans kept track of time.

happen to be phase-locked, something that is actually quite common in nature, and their synchrony will be maintained by their periodic cycle *automatically*.[16]

Even if you have to phase-lock the clocks, their synchrony is guaranteed by their constant period and you do not have to phase-lock all of them at the same time, which allows you to avoid the multiple-activation problem. The phase-lock relation among these clocks is genuine because it fully depends on their phase and period. However, the synchrony imposed on interval clocks is fully dependent upon when they are activated by a user and their synchrony is, therefore, not a genuine relation among them.

What about trying to build a set of interval clocks with a fixed relationship? This is the case in which different interval clocks turn into one contrived interval clock. Suppose you have a source of water that flows to a channel at a constant rate. Once this channel is full, say after a minute, it empties its content into a container that fills up every hour. The channel keeps filling up and emptying out its content. How, one might ask, is this different from the fixed relationship of the minute hand and the hour hand of an analog periodic clock? The difference is a fundamental one: this water clock is a contrived interval clock that, unlike an analog periodic clock, offers no advantages as a timekeeper because of the following reasons.

First, once the hour container is full, it spills. So this is really a *one-hour* interval clock. Suppose that there are other containers that accurately measure two, three, four hours and so on. There is a limit with respect to how many hours these clocks will measure. Moreover, because of property c) of interval clocks, once we get to the biggest container, say a gigantic pool of water, we will not be able to distinguish one hour from two or more hours. The easiest way to deal with this problem is to *empty* the container of the hour clock every hour (every time it fills up) and then keep track of the hours by counting them. But then, it is the *user* of the clock who is keeping track of the hours. There is no fixed relationship between the hour container (the hour clock) and larger-scaled ones, like the big pool of water. Their relationship depends on how the user keeps track of these containers. Unlike a set of periodic clocks, which keep their phase

[16] These synchronous phase-locked oscillatory processes are ubiquitous in nature (see S. Strogatz, 2003, and references therein). Strogatz claims that synchrony between oscillators is inevitable, based on the mathematics of coupled oscillators. Just to give a sense of the variety and scope of oscillatory processes that 'phase-lock,' here are some of the cases of synchrony that Strogatz analyzes: synchronal flashing of fireflies, satellite orbits, the chirping of crickets, insect outbreaks, human sleep, and superconductors.

relationship fixed in a reliable and mechanical fashion, a set of interval clocks require constant *monitoring by their user*.

What about the relation between the hour container and the minute one? Again, there is no fixed relationship between these containers: their relation depends on how the user keeps track of their contents. It is true that if the minute interval clock accelerates its rate, then the hour interval clock's rate will accelerate accordingly. But what we have in this case is a single unreliable hour clock that accumulates water at a fluctuating rate. The amounts of water that the minute clock pours are only smaller intervals of this unreliable hour clock. If we think of the minute interval clock as an independent clock, then monitoring will be required in order to keep track of every time it empties its content. It will be the user of this contrived hour clock who keeps track of the minute clock. Thus, there is no reliable and mechanical relationship among interval clocks and all their possible relations depend on the user.

The third disadvantage of interval clocks is their lack of sensitivity to periodicity. This is an important shortcoming when the environment has a vast number of periodic events happening regularly, as is the case on earth. If at sunset there is apple pie for free at the coffee shop, knowing when to run to the coffee shop is quite easy if one has a periodic clock of 24 hours that could be used as an alarm clock: one only needs to phase-lock the clock to 'sunset' and then, mechanically and automatically, one will always get on time for free apple pie.[17] But if one has an interval clock of 24 hours, then one needs to activate this clock after each apple pie meal, and worry whether one activated it on time for the next day. The clock by itself would go ahead and end its interval. If the user is not careful, and activates the clock 2 hours after having apple pie, then she will be two hours late for apple pie the next day.

Finally, a disadvantage that is implicit in what I have just said, particularly concerning the second and third disadvantages, is that interval clocks are much more user dependent than periodic clocks. The automatic processes of periodic clocks (e.g., recurrent cycles, fixed-phase relationships and phase locking) are an incredibly important basis for *automatic behavior*, sensitive to periodicity and temporal information. Interval clocks literally run out of whatever media instantiates them and

---

[17] This is assuming that 'sunset' always happens at the same time, which is not the case in many latitudes on earth. But for the sake of the example this assumption is not problematic. However, as I will explain, periodic clocks can be re-phased so that they are always phase-locked to a regular event.

stop, but periodic clocks are, as it were, always running. This is a crucial advantage of periodic clocks that creatures endowed with a circadian clock, which are basically all living creatures on earth, exploit constantly, as I will explain in the next section.

To conclude, I will summarize the four disadvantages of interval clocks I have discussed and then proceed, in the next subsection, to describe how periodic clocks may interface with interval clocks, thus allowing for a hybrid or mixed timekeeper. The disadvantages of interval clocks are:

1. There is always a scale/resolution trade-off.
2. They cannot be related to one another in a reliable mechanical way.
3. They lack sensitivity to periodicity.
4. They lack automatic, self generated processes that can mechanize behavior and are correspondingly user dependent.

2.4.3. *Hybrid and Semi-Hybrid Clocks*

One way to overcome the inconveniences of periodic and interval clocks is by designing a hybrid or semi-hybrid clock. An important distinction needs to be highlighted between a hybrid clock and the *use* or *interpretation* of a periodic or interval clock according to information external to, or not encoded by, the clock. For instance, dating based on tree rings and ice core layers is, in a sense, a hybrid technique for timekeeping because scientists have to *interpret* the accumulation of the layers produced by these periodic clocks as constituents of an *interval* clock. But scientists have other information, such as calendars, to compare tree rings and ice cores in order to determine intervals of time in terms of years. The dating of events depends not only on calendars, but also on other technical calculations, such as those based on Carbon 14. If a scientist were given a tree trunk or an ice core sample and nothing else, dating the rings or layers of ice would be quite difficult.

However, because they are *registering* year intervals, trees and ice cores are what I shall call 'semi-hybrid' clocks. The mechanism that produces the annual cycle encoded by trees and ice cores is the rotation of the earth around the sun, which is a periodic clock. But the fact that trees and ice cores register this period of time by *growing* makes them semi-hybrid clocks in the following sense: it is not *accumulating at a constant rate* that makes these clocks accurate, because trees and ice cores need not have a constant growth or accumulation rate in order to accurately encode periods of one year. They only need to grow, and it does not matter if their growth rate fluctuates. So, they are semi-hybrid clocks because they have a register for periods of one year, instantiated as wood and ice, and a *reliable periodic clock* that marks these periods (i.e., the rotation of the earth around the sun).

Notice that the name 'semi-hybrid' is adequate because it would be a mistake to attribute any of the properties of interval clocks to tree rings and ice cores ('semi-hybrid' means that the mechanism is a combination of a clock and a register, rather than a combination of *two clocks*). As mentioned, tree rings and ice cores do not comply with property a) of interval clocks (their constant rate). But they do not comply with other properties too, such as c), because tree rings and ice core layers need not emulate an interval: having a register, they can cover very large as well as very small scales and still have excellent resolution for periods of about one year. In contrast, sand clocks, and interval clocks in general, emulate intervals of a specific scale and have the scale/resolution problem. Thus, tree rings and ice cores are semi-hybrid clocks because they are used as *registers* of the repeating cycles a periodic clock, and this eliminates disadvantages of interval clocks. However, they are not the combination of an interval and a periodic clock.

A truly hybrid clock combines a periodic clock with an interval clock, and each of them satisfies their respective properties. The ideal combination is a periodic clock that determines the rate at which an interval clock accumulates. If one represents equal accumulation segments in terms of numbers, then one would obtain a hybrid clock with an oscillatory mechanism that feeds information to a counter. Notice that numerals, which make the accumulation process a *counter*, are representations imposed on the accumulation process, and they are not necessarily constituent parts of the clock. This is the most accurate way we have come up with to measure time.[18] Examples of hybrid clocks are the atomic clock and the stopwatch, which I describe in section 2.6. It is important to distinguish carefully between the *register, or memory for events*, such as a counter or a calendar, and the *clocks*, which have their own reliability constraints. With respect to the principles governing the combination of clocks, there are four possible ways in which clocks can interface:

[18] The atomic clock counts the oscillations of a cesium atom. The optical clock is based on the same principle and it is even more accurate than the atomic clock. Our use of numerals allows us to simply count oscillations, but if we lacked numerals, an interval clock that grows at a constant rate marked by these oscillations would do a similar job.

1. The combination of two or more periodic clocks with the same or different periods.
2. The combination of two or more interval clocks.
3. The combination of an interval clock with a periodic clock, where the periodic clock determines the interval clock's constant rate.
4. The combination of a periodic clock with an interval clock, where the interval clock measures the durations between phases of the periodic clock.

I argued that (1) is a significant advantage of periodic clocks. I also argued, against (2), that there are no genuine relations among interval clocks. (3) and (4) are interesting cases of clock-combinations, and they are defined by the following principles:

> **Truly Hybrid**: In the case of the combination of an interval clock with a periodic clock, where the periodic clock determines the interval clock's constant rate, the resulting hybrid clock will be an *interval clock*, which means that it will comply with the characteristics of interval clocks and the periodic clock will only be relevant to keeping its rate constant.

> **Periodic Segmentation**: In the case of the combination of a periodic clock with an interval clock, where the interval clock measures the durations between phases of the periodic clock, there is no *genuine combination of clocks*. The periodic clock's phases are not *determined* by intervals. Rather, the interval clock is *used to measure* the duration of segments of the periodic clock's cycle.

As mentioned, the atomic clock and the stopwatch are cases of hybrid clocks that are truly hybrid, as my discussion on the stopwatch in section 2.6. illustrates. Computations on periodic clocks, performed for the purpose of calculating *intervals*, comply with 'periodic segmentation.' My characterization of periodic and interval clocks emphasizes their reliability as timekeepers. But one may ask, why devote so much attention to the constraints on the reliability of these clocks? What is the philosophical interest of this discussion and why is it useful for a general theory of time representation in psychology? I have three responses to these questions.

First, it is useful to have a theoretical account of how registers of time are capable of satisfying the uniformity of time and units of time constraints for accurate time measurements (i.e., time must flow at a constant rate and the units of time must be invariant). The first four sections of this chapter provide this theoretical framework, which will serve as the basis for my explanation of how the circadian clock and the stopwatch can be used for epistemic purposes by animals and humans.

Second, the relationship between the clocks and the processes they emulate (periods or intervals) is important to understand how mechanical devices and living organisms are capable of registering time with great accuracy. Mechanical explanations are important for naturalistic or scientifically informed accounts of mental representation. Thus, this discussion is going to be very relevant to understand time representation in animals and humans based on scientific findings, without invoking complex philosophical issues, such as the nature of the self.

Finally, reliability is a central notion in naturalistic theories of justified belief. If one wants to find the source of our immediate beliefs about the duration of events, and explain what could make them *justified* beliefs, one could appeal to the fact that they are produced by a reliable belief forming process: the processes that instantiate the clocks I am about to describe. I will come back to this issue. For now, it suffices to say that the interface between periodic and interval clocks demonstrates that they are different forms of measuring time accurately. They can be combined precisely because they are reliable and comply with the basic constraints for time measurement. I shall now proceed to explain the circadian clock and its relevance for the representation of time in living organisms.

2.5. The Circadian Clock

In this section, I describe the circadian clock and analyze its most important features. I first provide a brief introduction to the findings on biological rhythms, including *ultradian* and *infradian* biological cycles. In subsection 2.5.1., I discuss entrainment, in the context of experimental evidence concerning the circadian clock. Finally, subsection 2.5.2. is an assessment of two types of representations based on the circadian clock: sun-compass navigation and memory for time of occurrence.

The circadian clock is a *periodic clock* with a cycle of about 24 hours, which emulates the rotation of the earth on its own axis. Although there are other oscillatory processes that could be considered as biological clocks, such as those involved in lunar, seasonal and annual rhythms, the circadian clock (and its corresponding daily rhythm) is by far the most important, and most studied, biological clock. Scientists have found that plants, fungi, bacteria and animals have circadian clocks, which regulate a vast variety of rhythmic behaviors.

The fact that circadian clocks have been scientifically identified in organisms as different as bacteria and plants is remarkable. There are fascinating aspects of circadian rhythms. Perhaps the most amazing one is that the circadian clock regulates very important biorhythms, beginning with genetics, with marked effects at the organism level. This explains why the circadian clock is often referred to as the 'master clock.' For instance, the circadian clocks of different animals influence niche formation and biodiversity in ecosystems by partitioning animals into diurnal and nocturnal. From gene transcription to sleep cycles, the circadian clock's impact on an organism and its environment is truly incredible. As Panda and Hogenesch (1998) explain:

> Circadian modulation of gene transcription translates into rhythmic activity of their protein products, which in turn generates rhythmic metabolic flux in different tissues and rhythmic behavior and physiology at the organism level, thereby enabling the animal to adapt to diurnal changes in its environment (Panda and Hogenesch, 1998, 375).

There are many other interesting issues about the circadian clock that have been discussed in the vast literature generated by its multi-disciplinary study. However, I shall focus only on the most general aspects of the circadian clock that make it a reliable periodic clock, and on its relevance for the *representation* of time in organisms. But, before addressing the characteristics of the circadian clock, I will first introduce some technical terminology concerning the distinction between the circadian clock and other biological oscillatory mechanisms, and their corresponding biorhythms.

The period of the cycle of the circadian clock is so important that other biorhythms are defined in terms of the circadian rhythm ('circadian' literally means *about one day*: 24 hours). *Ultradian* are those rhythms and oscillatory processes that occur more frequently than every 24 hours. These ultradian rhythms may determine a biological cycle within one of the rhythmic functions regulated by the circadian clock, such as sleep. For example, Daniel F. Kripke (1973) has identified a 90 to 120-minute ultradian rhythm that regulates the alternation between rapid and non-rapid eye movement sleep. This ultradian oscillatory process has also been related to waking gastric activity and brain functions during wakefulness, related to fluctuations in attention (D. F. Kripke, 1972; J. F. Hiatt and D. F. Kripke, 1975).

Biological rhythms that occur less frequently than every 24 hours are called *infradian*. Some examples of infradian cycles are the menstrual, lunar and circannual cycles ('circannual' means *about one year*). Circannual cycles have been identified in hibernating mammals, such as ground squirrels (G. R. Michener, 1979), and woodchucks or groundhogs (P. W. Concannon et al., 1992). Lunar cycles have been identified in sea creatures, which have lunar clocks that allow them to synchronize their biological rhythms with tidal changes.

A fascinating example of an animal with a lunar clock is *Clunio marinus*, a type of fly. The larvae of *Clunio marinus* are extremely sensitive to moonlight. G. Fleissner et al. (2008) found that the *ocelli* (photoreceptor organs that could be considered primitive eyes) of the *Clunio marinus* larvae change their shielding pigment transparency according to a lunar-rhythmic cycle. This allows the larvae to adjust to changes in light intensity, which depend on various factors including tidal changes. The fact that the shielding pigment transparency of the larval *ocelli* changes according to a lunar cycle strongly suggests that *Clunio marinus* has a lunar clock that regulates these changes.

The list of ultradian and infradian rhythms (and the ultradian and infradian clocks that produce them) keeps increasing. Animals are finely tuned to periodic changes in their environment and have developed clocks to track important periodic events. Chronobiology will certainly produce more exciting and unexpected results concerning ultradian and infradian rhythms. However, the circadian clock is the most important biological clock for two reasons.

First, its biological basis and functions have been identified in animals and plants, and studies reveal that the circadian clock regulates a vast amount of rhythmic behavior. And second, which is the most relevant issue for the topic of time representation, the circadian clock of animals with nervous systems plays an important role in producing representations of the environment used in navigation and anticipatory behavior. I proceed to explain the main characteristics of the circadian clock—relating them to the general characteristics of periodic clocks—and to justify these claims.

2.5.1. *Entrainment*

Entrainment occurs in all periodic biological clocks, and it correlates with the capacity of periodic clocks to have fixed-phase relationships, including synchrony. If there is a fixed relationship between an *environmental* periodic clock and a periodic *biological clock with the same period*, then their cycles are phase-locked. The circadian clock's cycle is phase-locked with the cycle of the rotation of the earth on its own axis. Thus, the circadian clock (or any other periodic biological clock) is *entrained* by maintaining a fixed phase relationship to an environmental cycle. If a biological clock is entrained, then there is another periodic event to which its cycle is phase-related. This enables organisms to optimize their behavior with respect to cyclical environmental events and also with respect to the cyclical behavior of other organisms.

A periodic clock, or an oscillator that influences another oscillator is called the *master oscillator* and the one that changes as a result of such influence is called the *slave oscillator*. In the case of the circadian clock, the master oscillator is the period of 24 hours in which the earth rotates on its own axis. This periodic clock determines the cycle and phases of the circadian clock, which is, in this case, a slave oscillator. However, within a particular organism, the circadian clock determines the cycles and phases of many other oscillators, and this is why the circadian clock is also known as the master clock, because in those cases it is the master oscillator. Entrainment is the relationship that holds between a master and a slave oscillator: the slave oscillator is entrained by certain *cues* from the master oscillator, just like the circadian clock is entrained by certain cues from the environment, usually light intensity.

But what exactly is the influence that the master oscillator has on the slave oscillator? By maintaining a fixed phase relationship, changes in the master oscillator's cycle re-set the phases of the slave oscillator. These changes are cues that trigger a shift in the phases of the entrained periodic clock. Suppose that you isolate an entrained periodic clock from the cues of the master oscillator that are relevant for re-setting its phases and cycle. Then the entrained periodic clock is said to be "free running," and although it can still reproduce its cycle *accurately and reliably* (in the case of the circadian clock, every 24 hours) its phase relationship with the master oscillator has been severed.

The circadian clock of plants illustrates this point. Since plants depend mostly on sunlight for their survival, it is no surprise that their circadian clock can be *re-set* by manipulating light cues. However, they reproduce their circadian rhythm even in the absence of light. Andrew J. Millar (2003) explains:

> Circadian rhythms in plants are relatively robust, as they are maintained both in constant light of high fluence rates and in darkness. Plant circadian clocks exhibit the expected modes of photoentrainment, including period modulation by ambient light and phase resetting by brief light pulses (Millar, 2003, 217).

The plant's circadian clock is accurate even in the absence of light, when it is free running. But its period can be modulated by light intensity and its phases are re-set by light cues. When the circadian clock is free running, there is a *phase drift*, which indicates the significance of the mismatch between the circadian clock's cycle and the external cycle that entrains it. If the phase drift is very significant, then the circadian clock will have to transiently accelerate or slow down its pace until it can re-establish the phase relationship with the external cycle again. This holds for all circadian clocks, as is illustrated by the circadian clocks of cockroaches and hamsters (see Gallistel, 1990, 229–231).

Entrainment has an enormous significance for the survival of organisms that have biological clocks. All the advantages of periodic clocks are manifest in entrainment (e.g., mechanically reproducible relationships to other periodic clocks, sensitivity to periodicity and automatic behavior). Entrainment is responsible for keeping organisms in tune with their environment, including other animals and conspecifics. For example, certain phases of the clock could be phase-related to the periodic presence of prey, which is an identical situation to the one I described in section 2.4.2. (the apple pie example). One of the most dramatic examples of a set of synchronized phase-locked biological clocks in nature is the existence of swarms of insects or birds that reproduce or feed at the same time.

The automatic and reliable mechanisms that underlie entrainment are present in all organisms. Circadian rhythms are ubiquitous. I mentioned plants and animals, but I could have talked instead of bacteria and fungi.[19] Nervous systems and brains, however, provide the possibility of using circadian clocks to register information about the environment, which would constitute a *semi-hybrid clock*, composed of a periodic clock and a memory-register.[20]

Plants have photoreceptors and a complex biochemistry that underlies their circadian clock. But plants do not seem to have a semi-hybrid clock. They are best characterized as having a *natural periodic clock*. As we go up the scale of neural complexity, organisms with nervous systems are able to encode information in memory. In the case of vertebrate animals, who possess the most complex nervous system (composed of a central and a peripheral nervous system) the registration of information in the brain

[19] See G. S. Golden and S. R. Canales (2003) for circadian rhythms in cyanobacteria and J. C. Dunlap and J. J. Loros (2006) for circadian rhythms in fungi.

[20] I describe semi-hybrid clocks in section 2.4.3. of this chapter.

creates a very sophisticated *semi-hybrid natural clock,* constituted by the circadian clock and the brain's memory components.

However, even the small network and set of ganglia that constitute an insect's nervous system are powerful enough to produce an amazing semi-hybrid clock, with enormous repercussions for the animal's behavior. I proceed to describe the representations encoded by animals with semi-hybrid clocks.[21]

2.5.2. *Circadian Clock Representations*

A great achievement of modern biology was the identification of the suprachiasmatic nucleus (SCN) as the locus of the master circadian clock in mammals. The origin of this important discovery is the work of Carl Richter (1967), who demonstrated that lesions to the frontal part of the hypothalamus of rats eliminated rhythmic behaviors. The identification of the SCN as the locus of the circadian clock is due to research by Stephan and Zucker (1972), Moore and Eichler (1973) and Ralph, Foster, Davis and Menaker (1990).[22]

The circadian clock synchronizes central and peripheral oscillations. The SCN is located in the hypothalamus, above the brain stem. The neurons of the hypothalamus and the brain stem regulate neurological functions necessary for survival, and the biochemical reactions that underlie these vital functions are kept in synchrony by the SCN. The circadian clock of other creatures is, like the SCN, also responsible for regulating vital functions. However, creatures with nervous systems have a crucial advantage for registering time based on their circadian clocks, because other neural networks of their brains can utilize the information contained in the SCN.

In the case of creatures that are not mammals, their ganglia or brains can utilize information from their circadian clock, whichever way it is instantiated. This generates a neural semi-hybrid clock as follows: in mammals, the SCN instantiates the periodic circadian clock, and other regions of the brain instantiate a register for information concerning the SCN (and similarly with other neural systems). The SCN works automatically and the other neural networks keep track of its activity.

---

[21] What about trees? I said that they are semi-hybrid clocks, but this is because *we use them* as semi-hybrid clocks. Clearly trees and ice-cores cannot *use* the information contained in their registers for annual cycles. The difference is that animals with nervous systems *can* use the information stored in their brains.

[22] See J. Aschoff (1965) for studies on human circadian rhythms.

Thus, other specialized areas of the brain may utilize information from the circadian clock to compute *duration*, rather than periodicity, but the circadian clock will remain a periodic clock. This is consistent with my account of semi-hybrid clocks and the segmentation of their cycles. But in this case, there is no interval clock measuring phases. Instead, computations on information stored from the circadian clock are used to determine the length of intervals. I will present two cases, tested experimentally, of how information obtained from the circadian clock can be utilized by other specialized areas of the brain for representational purposes. One of them concerns navigation and the other the registration of time of occurrence, or the attribution of a temporal tag to an event.

2.5.2.1. *Navigation and the Solar Ephemeris Function*

It may seem counterintuitive to use a self-sustaining and automatic mechanism like the circadian clock for sun-compass navigation, which depends entirely on environmental information. However, when other specialized areas of the brain properly interpret information obtained from the circadian clock, it can become a powerful component of a navigation system. It would be inaccurate to say that the circadian clock is used *as a clock* in sun-compass navigation. What is being used is information concerning certain phases of the clock that get interpreted as *spatial* information concerning the angle and position of the sun. Insects and birds can perform such navigational computations. The experimental challenge is to confirm that they are indeed using information from the circadian clock.

The idea is that sun-compass navigation in animals might exploit phase information from their circadian clocks to compute the position and angle of the sun in order to determine their orientation by fixing which direction is north. This requires representations of the environment, including spatial interpretations of the phases of the circadian clock. How to test this hypothesis experimentally? The easiest way is to design an experiment that tests whether changes in the phases of the clock correlate with predictable changes in orientation during navigation. As Gallistel says, one way to demonstrate the existence of an internal ephemeris function "that gives the sun's azimuthal position as a function of the time indicated by an animal's endogenous circadian clock is to put the endogenous clock out of phase with the local day-night cycle (so-called clock-shifting experiment)" (1990, 81).

The experimental evidence confirms that animals use their circadian clock to compute the ephemeris function. M. Renner (1960) trained bees in Long Island to fly to a feeding station north of their hive at a compass bearing of 315 degrees. He then packed the bees at night and flew them to Davis, California. Gallistel explains:

> The time difference between Long Island and Davis is a little more than three hours, which means that the azimuthal angle of the sun at Davis is on average about 45° behind its azimuthal angle on Long Island. The bees from Long Island were jet lagged when their hive was opened at Davis; their endogenous clocks were roughly 3 hours ahead of the day-night cycle at Davis. When a bee whose internal clock is 3 hours ahead of local time tries to use the sun to set a northwesterly course, it will in fact set a westerly course; it will orient approximately to 315° – 45° = 270°. This is what the bees did at Davis, proving that their ability to steer by the sun depends on an endogenous ephemeris function linked to their endogenous circadian rhythm (Gallistel, 1990, 81).

Similar results have been confirmed in experiments on ants (R. Wehner and R. Müller, 1993) and homing pigeons (S. T. Emlen, 1975). It is remarkable how information from the neurons that instantiate the circadian clock is transformed by other regions of the brain into spatial information for navigation. This requires specific metric information that maps onto features of the environment. The brain uses circadian clock information to compute the ephemeris function, which is the result of a combination of representations with *metric structure* (a topic that I will examine in the next chapter). However, the following example of circadian clock representations is more pertinent to the topic of temporal representation because it is a case of what I described as a semi-hybrid clock, instead of a navigation system-component. It also illustrates how a specific time is attributed to objects and events, based on the clock's representations.

2.5.2.2. *Time of Occurrence*

Experiments on the capacity of animals to register the time of occurrence of events were originally based on the observation that insects, birds and mammals *anticipate* the availability of food within a specific time range: not too early, because that would only be a waste of time, but not too close to the arrival of conspecifics and other competitors before food is available, because in that case, anticipating the availability of food would not be advantageous. This kind of anticipatory behavior has been confirmed in bees, birds, and rats.[23] The question is, how are these animals anticipating events and registering their time of occurrence?

---

[23] See C. R. Gallistel (1990, 243–286) and references therein.

There are two possible explanations for anticipatory behavior, based on the characteristics of the circadian clock. The first is that a particular event is *associated* with a phase of the circadian clock. The memory for the time of occurrence would be phase-dependent (also known as state-dependent) because it would only be available to the animal when it is in that phase of the cycle. There is no *register* for times correlated with phases. Rather, these are only phases of the circadian cycle that are *identified with* some kind of marker, which indicates to the animal that food is about to arrive. The animal learns how to anticipate events by associating them with phases of the circadian cycle through repetition: the more frequent the occurrence of an event at a phase, the stronger the association. In order for the marker to activate and produce anticipatory behavior, the animal's circadian clock has to be at the phase of the cycle that correlates with it. This is a model for time of occurrence that relies exclusively on the circadian clock and its capacity to *phase-relate* periodic events to its cycle through entrainment.

To illustrate this point, imagine that the circadian clock of the bee (or bird or rat) is a kind of Ferris wheel with only one gondola. The animal is "sitting" in the gondola as the wheel spins. The wheel has a cycle of 24 hours and the gondola where the animal is sitting has a screen that tells the animal what to do. When there is an important event (like feeding) the screen displays the message 'food' and registers the event by marking the phase of the cycle at which the event happened. Every time the wheel is at this phase, it displays the message 'food', but the animal has no access to that information at any other time. Since the period of the cycle is 24 hours, the message produces behavior in the animal every 24 hours. Adjustments to changes in time are accommodated by entrainment. Anticipatory behavior can be explained by the capacity of the circadian clock of animals to adjust to phases that *optimize* feeding. Call this model the *phase-based* model.

The alternative model assumes a semi-hybrid clock, constituted by a clock (the circadian clock) and a register or memory *for* times (i.e., an episodic-like memory), rather than a phase-based mechanism. The main difference from the previous model is that times of occurrence are registered in memory, along with other information, and thus the animal has access to this information at *any* time (not only at specific times that correlate with a phase). This does not mean that phase information is no longer relevant: mismatches between the phases of the clock and periodic environmental events caused by phase-shift will disorient the animal, which will have to re-entrain the clock. However, the information concerning phases and events will be available to the animal at any time. Call this model the *memory for time* model.

The *phase-based* model is attractive because it assumes a simpler, more economic (in terms of brain power), and self-sustaining or automatic mechanism: the circadian clock. Researchers on biological rhythms tend to favor this kind of model because of the overwhelming evidence in support of entrainment in circadian clocks, which explains how animals can learn to anticipate food availability. In principle, plants and rats might be using the same phase-based model. In the case of plants, phase-based information "tells" the plant how to balance water, when to change its position and orientation in order to absorb more light, and how to optimize energy consumption within the 24 hours cycle of the circadian clock.

In the case of the rat, things are more complicated, partly because the nervous system of the rat is much more powerful than the set of sensors of the plant, but the model might be the same. Phased-based information from the circadian clock might alert the rat to all sorts of things: when to get food, wake up, sleep, increase levels of activity, etc. Why should we introduce the assumption that memory and extra computations are needed to provide an accurate model of the anticipatory behavior of animals? The answer is simple: because the experimental data confirm the memory for time model in insects and animals and show that it is indispensible to explain anticipatory behavior.

Evidence in favor of the memory for time model comes from experiments on anticipatory feeding behavior in bees. A series of experiments conducted by R. Kolterman in 1971, demonstrates that bees can distinguish 19 *different* times of the day, which were correlated with different odors. Crucially, they can learn information about times after only a single day's training.[24] This shows that what is relevant for anticipatory behavior is not the periodicity of events, but rather the time at which these events occurred, as registered in the episodic-like memory of animals. The clock serves as the basis for successfully measuring times, but it is the register that keeps track of times and durations. The irrelevance of periodicity found by Kolterman is strong evidence in favor of the memory for time model. I shall emphasize that phase information is crucial for this model: these representations stored in memory are *readings* of the circadian clock at a particular phase.[25] This is demonstrated by the fact that the

[24] See Gallistel (1990) and references therein for details and further discussion on Kolterman's experiments.

[25] In the next chapter, I explain how these readings are computations on the outputs of the clock.

behavior of the bees the next day correlated with the phases of the circadian cycle and not with the one-hour period of the training schedule, because their search behavior *did not repeat every hour*. Rather, the bees searched *only at the phase* of the cycle in which they experienced the relevant stimulus (geraniol). However, the finding that bees learn how to time their search in *a single day* according to feeding time, rather than the periodicity of feeding recurrence (one hour), plus the finding that they can distinguish 19 different times of the day (as individual times relevant for behavior), challenges the phase-based model on two fronts.

First, if animals are using their circadian clock to anticipate events, then it is difficult to explain how they can learn information about times within a 24 hours period without using a register for time of occurrence. If periodicity and entrainment are the main mechanisms underlying anticipatory behavior, then the prediction of the phase-based model is that the circadian clock must go through *at least* one cycle of 24 hours in order to mark certain phases as behaviorally significant. But the data from the Kolterman experiments prove this prediction wrong. Bees could learn schedules of only one-hour and distinguish as many as 19 different times within only one cycle of the circadian clock.

Second, the finding that bees can associate odors with times shows that there is information stored in a register, suggesting the existence of a computational semi-hybrid system, rather than a single periodic clock. Moreover, the attribution of a time to a specific odor satisfies a crucial constraint for mental representation because bees are attributing times to environmental *particulars* (i.e., the bee is attributing a time to a specific odor). Burge (2010) offers an account of mental representation that relies on this constraint because otherwise it is difficult to distinguish strictly mechanized behavior from *representation-based* action.[26] It seems that bees qualify as representation-based agents with episodic-like memory.

Similar findings, which give further support to this model, have been confirmed in the rat.[27] But the most dramatic example of episodic-like memory in animals comes from research on scrub jays, which I will briefly discuss in the next chapter, and on nonhuman primates. As William Roberts (2007) says, the findings on memory for times in scrub jays and

[26] Burge (2010) also says that the representations must be used by the animal at the organism level, rather than at the sub-personal modular unit-level. This is a more controversial constraint, but it is one that bees *also* satisfy in navigation. These two constraints explain why many neural clocks, unlike the circadian clock, do not qualify as mechanisms that produce temporal *representations*.

[27] For a detailed discussion on the evidence for the memory for time model in the rat see Gallistel, 1990, pp. 277–285.

nonhuman primates suggest that some animals can mentally time-travel not only into the past, but also into the future. These findings cover a vast amount of behavior involving planning and decision-making. Collectively, they confirm that a semi-hybrid clock, composed of the circadian clock and an episodic memory register, underlies the complex behavior of these animals. This is the most plausible explanation of anticipatory behavior. There is no need to invoke conceptual content, particularly concerning concepts such as 'cause' and 'self' to account for such complex behavior.

I discussed two types of circadian clock representations, and two models for anticipatory behavior: the phase-based and the memory for time models. It seems that only animals with brains (or a complex nervous system that supports a semi-hybrid timekeeper) are capable of representing circadian information for sun-compass navigation and time of occurrence. The evidence favors the memory for time model. But what kind of representations are these? How can they be characterized using current theories of mental content? How is it possible to define these representations without appealing to concepts concerning causality and the self? The main purpose of the next chapter is to address these questions. Before addressing them, however, I will describe the stopwatch, which is significantly different from the models examined above with regard to *memory*, *registration*, and *representation*.

2.6. The Stopwatch

The so-called 'stopwatch' is an *interval clock*. It is a biological interval clock, neurologically instantiated in the brain, with a short scale, ranging from seconds to minutes. Neuroscientists are investigating whether there might be different interval clocks that make possible the registration and production of sub- and supra-second intervals in different modalities (W. H. Meck, and C. Malapani, 2004). There is intense debate concerning the location of an interval timing mechanism for motor and non-motor activity, and the evidence favors the existence of a distributed network of neurons, rather than a single region of the brain (Y. Bhattacharjee, 2006).[28] This distribution distinguishes the neural correlates of interval timing from the neural correlates of the periodic circadian clock, which is located

[28] This evidence on the neural correlates of interval timing is based on experiments on humans. I will discuss evidence on neural mechanisms for interval timing in animals when I address the issue of the rate of the stopwatch.

in the hypothalamus, specifically in the SCN. For the sake of simplicity, I will focus on the characteristics of the stopwatch, assuming that it is a single mechanism. If it turns out that there is more than one stopwatch, that finding would not challenge what I am about to say because, as I will explain, all such timekeepers would behave like interval clocks.

Why would animals with circadian clocks and brains that allow the registration of information in memory evolve *another* clock for short intervals? This interval clock is not only very different from the periodic clocks just mentioned, but also seems to be distributed, rather than located in a specific region. The most plausible explanation, put forward by scientists, is that the stopwatch is deeply related to perceptual *attention* and is used to integrate information about duration coming from different sensorial sources.

The cross-modal nature of the information processed by the stopwatch fits well with its being a distributed network of neurons that can communicate with the specialized areas of the brain where the different modalities are located. This means that the stopwatch is likely to have evolved later than any periodic clock, almost certainly later than the circadian clock, which is ubiquitous throughout nature. The circadian clock seems to be a very old part of the brain and of the mechanisms that sustain life in general. The thesis of the recent evolution of the stopwatch receives further support from the significant amount of brainpower it requires, because it seems to be responsible for estimating intervals concerning attention-related tasks that require significant cognitive effort. In contrast, a substantial portion of the behavior controlled by the circadian clock is mechanized by entrainment and does not require attentive cognitive effort.

The recent evolution of the stopwatch can also be explained by the advantages that calculating intervals of seconds and minutes *without representing periodicity* offers to an animal.[29] F. B. Gill (1988) has shown that the hermit hummingbird's ability to forage efficiently (and even mate, in the case of the male) depends on its ability to represent temporal intervals with great precision. For the hummingbird, an interval clock is not only an advantage; it is actually *indispensable*. For instance, flowers do not fill cyclically because they are emptied by different animals at different times of the day that do not correspond to specific periods. However, flowers refill at a relatively constant rate. Birds must estimate the rate at

[29] For evidence on the capacity of animals to represent elapsed intervals with fixed interval schedules see J. Gibbon (1977) and P. Killeen (1975).

which they refill to feed efficiently. Emulating the rate at which flowers refill requires an interval clock for short durations. In an environment where events that are crucial for the survival of an animal occur aperiodically, having a stopwatch is a great advantage. A lot of the processes that occur in nature are periodic, and this is the reason why periodic clocks are present in all living creatures. But in an environmental niche, where there is constant competition, aperiodic events become commonplace.

Beside its probable recent evolution, however, what are the properties of the stopwatch? The stopwatch is a *hybrid clock* because it ultimately depends on the neural oscillations of the brain to keep its rate constant. But, for all intents and purposes, it is an interval clock, because it complies with the principle that I called 'truly hybrid,' as I am about to explain.[30] Moreover, all the models of the stopwatch offered in the literature describe it as an interval clock. If these models are correct, one should find all the characteristics of interval clocks in the stopwatch, just as one finds all the characteristics of a periodic clock in the circadian clock. Indeed, the stopwatch complies with all the characteristics of an interval clock.

Property a) of interval clocks, their constant rate, has been identified in the stopwatch by manipulating the biochemistry of the brain. Experiments have shown that certain substances accelerate the stopwatch's rate, creating the impression of time "slowing down;" they may also decelerate its rate, creating the opposite effect (Cheng, et al., 2006). Dopamine seems to be a relevant factor for the constancy of the stopwatch's rate. For instance, dopamine improves motor and non-motor interval timing. Dopaminergic antagonists slow down the stopwatch's rate while dopamine agonists speed it up.[31] A neurobiological cocktail seems to maintain the stopwatch's rate constant, which is the neural equivalent of having a good sand clock by getting everything right.

Property b) of interval clocks—that they are started by an activation event—corresponds to the activation and deactivation of the stopwatch, which is associated with perceptual attention. This means that the agent *activates* the clock by attending to a stimulus or set of stimuli and *deactivates* the clock by shifting attention or by decreasing the allocation of attention. An experiment by J. T. Coull et al. (2004), which tested non-motor interval timing, found that increases in attention in a timing task increased brain activity in specific areas of the brain. Meck and Malapani (2004) report the main results of this experiment as follows:

---

[30] See section 2.4.3. for discussion on this principle.

[31] See Meck and Malapani (2004) and references therein.

> The study made use of dynamically changing visual stimulus attributes (e.g., color and duration) that could be differentially attended to by the participants. Increasing attentional allocation to the temporal integration of stimulus duration selectively increased activity in a cortico-striatal network that included the pre-supplementary motor area, the right frontal operculum, and the right putamen. Conversely, increasing attention to the integration of the temporal variation in the color of the stimulus (rather than its duration) selectively increased activity in visual area V4. Thus, by parametrically increasing the attentional demands of the psychophysical task these researchers were able to identify the neural substrates of time and color perception (Meck and Malapani, 2004, 133).

The relevance of this finding to the topic of the activation and deactivation of the stopwatch through attention is that the stopwatch, which integrates the duration of stimuli, works similarly to a sense organ. The allocation of *attention* to the duration of stimuli serves as the activation event that starts the stopwatch, and it stops when the allocation of attention drops. There is no periodicity involved in this mechanism and thus, no possible way of modeling it in terms of entrainment or phase relationships. The activation and deactivation events of the stopwatch correspond exactly to the general property of interval clocks that makes them user dependent.

Property c) of interval clocks—the restrictions on the physical medium, including the inverse correlation between size and resolution—might be the main reason why psychologists needed to distinguish interval clocks from periodic clocks. The scalar property of interval timing that corresponds to property c) of interval clocks (experimentally demonstrated by Gibbon et al., 1984; Church et al., 1994; Malapani and Fairhurst, 2002) is described by Meck and Malapani (2004) as "one of the major hallmarks of interval timing." They further observe that the variability associated with the precision of the clock "grows in proportion to the length of the interval being timed" (Meck and Malapani, 2004, 135).

In other words, the degree of error in calculating intervals increases proportionally with the duration of the interval. The scale-resolution tradeoff of the stopwatch is the foundation of the scalar timing theory, developed originally by John Gibbon.[32] As mentioned, this tradeoff is a consequence of the physical constraints on the medium that instantiates

[32] There are alternatives to the scalar timing theory, but they all have to account for the scalar property that underlies Weber's law. The consensus is that there is no need to postulate several theories, because the scalar timing theory is sufficient to explain the data on interval timing. For a detailed introduction to scalar timing theory see W. H. Meck (2003).

the clock, i.e., its constant rate and its capacity to emulate an interval. I will explain aspects of the scalar timing theory (particularly the importance of Weber's law) in the next chapter, where I examine sensory-motor representations of time with metric structure. But it is important to highlight that property c), which is fundamental to characterize interval clocks, is equally fundamental to characterize the stopwatch and its interval timing computations.

Finally, property d) of interval clocks is that they only register or represent *segments* of an interval, not *phases* of a cycle. This has two consequences for the components of the interval timing mechanism. Gibbon et al. (1984) describe different components for interval timing: clock, memory and decision components. Strictly speaking, the clock component is the one that corresponds to the stopwatch, but since the characteristics that define interval clocks apply to interval timing in general, the three components can be considered as part of the stopwatch. Property d) implies that the memories stored from the stopwatch are different from the memories that determine time of occurrence, stored from the circadian clock, because there is no periodicity or phase information concerning durations. There are two models for circadian-clock based behavior: the phase-based model and the memory for time model. In the case of the stopwatch, the two competing models, which I am about to describe, *require* memory. I suggested that the memory for time model provides the best explanation of the data on time of occurrence in animals. If this is true, then humans and animals with brains that are complex enough to have different senses, have two ways of registering time:

1. A phase-dependent format for memorizing the outputs of the circadian clock that contains information concerning periods; and
2. A phase-independent format for memorizing the outputs of the stopwatch that contains information about durations independently of any cycle.[33]

The stopwatch represents an interval's duration without any information about *periodicity* or recurrence of events. But if there are at least two components of the stopwatch e.g., clock and memory, how exactly do they interact? Many neuroimaging studies have tried to dissect the contribution of these different components (See Meck and Malapani, 2004).

[33] See section 2.5.2.2. for discussion on the distinction between a phase-based model and a *phase-dependent memory for time model.*

The Coull, et al. (2004) experiment is one example of such studies. However, besides the effort to locate the neural correlates of the components of the stopwatch, researchers must also develop a model of how the stopwatch works in terms of the information it computes. Since the stopwatch is an interval clock, it is pertinent to ask, what kind of physical process instantiates it: an accumulation or a decay process? The two models of the stopwatch (the pacemaker or accumulator model, proposed originally by Gibbon in 1970, and Meck and Matell's striatal beat frequency [SBF] model) address this question. I proceed to describe these models.

The pacemaker or accumulator model, as its name indicates, characterizes the stopwatch as an interval clock instantiated by an accumulation process. This model theorizes that the activity of certain neurons generates pulses at a constant rate. These pulses are then registered and "accumulated" by other neurons. As Bhattacharjee (2006) says, this accumulation process happens "in the same way that a cup placed under a steadily dripping faucet accumulates drops of water" (2006, 597). Bhattacharjee continues:

> As the receiving neurons register more and more signals, the sense of time that has passed grows. Moreover, quantities of accumulated pulses corresponding to specific durations are recorded in long-term memory, allowing an individual to compare newly encountered time intervals to those previously experienced (Bhattacharjee, 2006, 597).

The pacemaker or accumulator model has been the most influential model of interval timing to this day. It explains fundamental issues, such as the scalar property mentioned above, and it also provides a framework for explaining data on the stopwatch by predicting behavior in experiments on interval timing. But Warren Meck (once a strong supporter of the accumulator model) and his collaborators have recently criticized this model. Meck's main criticism is that the accumulation process postulated by the pacemaker model is too simplistic and does not capture the complexity of neural activity, which does not accumulate linearly, as the model assumes.

Meck and Matell (2004) provide the SBF model as an alternative, which they claim has the same predictive and explanatory power of the accumulator model without its problematic assumptions concerning neural activation. The basic idea behind the SBF model is that low frequency oscillations are used by striatal neurons to learn and recognize patterns of synchronous activity across different neural regions. These patterns are then correlated with *intervals* and the stopwatch can calculate intervals by selecting a particular pattern.[34]

It is unclear whether the process of selection of a pattern of activation by striatal neurons should be considered as a decay process. It is clear that SBF dispenses with the accumulator component of the stopwatch. If it is not an accumulation process, then the best way to describe it is as a probabilistic process in which a set of possible patterns of activation suddenly collapses into a single pattern. This process could be considered as a decay process because it depends on the physical transformation of a medium (neural oscillations) into something else (a definite pattern correlated with an interval).

However it is conceptualized, it is important to mention that SBF has been strongly criticized by some neuroscientists, such as M. N. Shadlen, who has described the model as "pure fantasy," because of the ubiquitous nature of synchronous spikes in the cortex.[35] Neuroimaging and behavioral data will certainly generate more controversy. SBF is an innovative and plausible model that might eventually become the prevailing theoretical framework to explain how the stopwatch works. But, at the present moment, the only safe thing to say is that the stopwatch is an interval clock instantiated in the brain across different neural regions.[36]

To conclude, I have offered empirical evidence that demonstrates the existence of neurologically instantiated mechanisms that keep track of time. One of them, the circadian clock, computes periodic representations of time that are used for sun-compass navigation and for the registration in memory of the time of occurrence of periodic events. The other mechanism is the stopwatch, which computes representations of time exclusively in terms of duration, independently of periodicity. These mechanisms and their representations need to be classified in a broader account of mental representation that relies upon current views in the philosophy of mind and cognitive science. The main purpose of the next chapter is to provide a theoretical account of these temporal representations.

---

[34] For details see Meck and Matell (2004) and references therein.

[35] See Bhattacharjee (2006).

[36] Gallistel (1990) has also offered an alternative model for interval timing. Moreover, he has recently put forward a strong objection against both models of the stopwatch. I briefly comment on these issues in the next chapter because they concern computational and representational topics. For a different model of interval timing that emphasizes the role of noise in accumulation processes see Simen, et al. (2011).

But before proceeding, it is important to restate the philosophical significance of these findings. The philosophy of time greatly benefited from the developments in physics that led to the general theory of relativity. Einstein realized that in physical theory, it is crucial not to introduce concepts that are experimentally ungrounded or assumptions about the structure of space and time that one finds extremely intuitive without empirical support. Instead, he proceeded by making as few assumptions about such structure as possible, always specifying what was meant by each term used for measurements in spacetime.

To successfully measure and represent time in general, one needs to relate clocks to events. One does not succeed at performing this task by defining *conceptually* the units of time. Rather, as Einstein recommended, one picks out arbitrarily a process that specifies, based on its characteristics, units for measurement. The conceptual system that allows for the development of physical science is based on these definitions. Because these definitions give immediate physical knowledge and coordinate arbitrarily chosen objects or events with concepts of measurement, Reichenbach called them 'coordinative definitions' (1958, 14). Thus, a specific physical process must be chosen as a clock in order to conduct time measurements. I described two different processes that qualify as clocks (periodic and interval) and provided evidence that shows that the brain instantiates a periodic and an interval clock. But in order to count as *reliable* clocks, two constraints must be met.

The temporal interval is the most fundamental geometric object that mathematicians and physicists use to study the structure of spacetime (See R. Geroch, 1978). The fundamental metric aspects of intervals are analyzed in the next chapter. The importance of this chapter is that it addresses the two main constraints on the reliability of clocks: the uniformity of time constraint (constant periodicity or rate) and the constancy of units of time constraint (congruous phases and segments of intervals), which are crucial requirements for accurate time measurements. I explained how the circadian clock and the stopwatch satisfy these constraints.

Reliability is a very important concept in epistemology, or the theory of knowledge. According to an influential theory of justification, a justified belief is one that is produced by a reliable process (See for instance A. I. Goldman, 1992). Reliable processes that form immediate beliefs (i.e., beliefs that are not dependent on the content of other beliefs, such as perceptual beliefs) lead to successful behavior. This is going to be very relevant to explain the cognitive significance of the reliable outputs of the clocks for belief formation and time representation, which do not require necessarily conceptual knowledge.

However, on the standard interpretation of what a belief is, representations and mental content are fundamental. One needs representations with content to have propositional attitudes, such as beliefs. I showed that the circadian clock and the stopwatch are reliable processes that could in principle lead to immediate justified beliefs about duration (at least in the case of humans), but I have not shown decisively that these processes produce mental representations. In the next chapter, I argue that the clocks produce mental representations and explain what type of representations they are.

CHAPTER THREE

SENSORY-MOTOR REPRESENTATIONS OF TIME, THE OUTPUTS OF THE CLOCKS AND THE TWO CONSTRAINTS ON MOTOR TIME COORDINATION

In the previous chapter, I categorized timing mechanisms into periodic and interval clocks. I explained why the circadian clock is a reliable periodic clock and why the stopwatch is a reliable interval clock. In this chapter, I address questions concerning the representational outputs of these clocks, e.g., what criteria they must satisfy to be considered representations, what kind of representations are they and what kind of information they contain? I answer these questions with a philosophical proposal concerning temporal sensory-motor representation, paying close attention to the experimental evidence.

I explained how the evidence on the anticipatory behavior of bees and hummingbirds for specific odors and flower replenishing rates satisfies the criterion that, according to Burge (2010), any legitimate mental representation must satisfy: representations produced by the clocks must attribute temporal sensitivities to environmental particulars at the organism level. These are not linguistic or conceptual representations. How, then, should we characterize the representations that are the outputs of the clocks?

A very important property of the outputs of the clocks is that they are representations with *metric* structure. In this chapter, I define metric structure and explain how such structure allows for the cognitive integration of the outputs of the clocks with other metrically structured representations. Understanding how this type of metrically structured cognitive integration occurs is fundamental to appreciate the important role that the clocks play within the sensory-motor system. I argue that circadian and stopwatch clocks are two independent systems for temporal representation whose outputs are crucial for motor coordination and action.

The structure of this chapter is as follows. Section 3.1. is an assessment of issues of representation and isomorphism. I explain why the temporal representations of the circadian clock and the stopwatch need to be understood in terms of isomorphism, and describe the different properties of the isomorphic representations of periods and intervals. This isomorphism allows animals and humans to rely upon the two constraints for accurate time measurements, based on the reliability of the clocks, as constraints on successful and precise motor coordination.

In section 3.2., I focus on considerations concerning the structure of these outputs, and I argue that the structure that frames these representations is metric, i.e., it preserves information about magnitudes. I review the relevant experimental evidence on the metric structure of the outputs of the clocks and discuss it in two subsections, one of them devoted to the circadian clock and the other to the stopwatch. I then explain the importance of the metric features of these representations, such as their relation to Weber's law.

Finally, in section 3.3., I argue that the best way to account for the metric structure of the outputs of the clocks and their isomorphism with respect to periods and intervals is by characterizing these outputs as *analog* representations. I present five criteria for defining analog representation and demonstrate that the outputs of the clocks satisfy all these criteria. Indeed, this section shows that the outputs of the clocks are *paradigmatic* cases of analog representation.

3.1. Representation and Isomorphism

John Heil (2005) says that 'disposition' "is a term of art: you can define dispositions as you please."[1] I believe that the same is true about the term 'representation.' Heil also says, however, that some ways of defining a term are more felicitous than others. Felicitousness depends ultimately on capturing the specifics of a particular case. Defining artistic or scientific representations poses specific challenges that are very different from the challenges one faces in defining mental representations. But even if one focuses exclusively on mental representations, there are different types of representations that generate their own definitional challenges.

The best way to capture the specific characteristics of the representations that I analyze in this chapter, namely the representations of time produced by the circadian clock and the stopwatch, is by taking what J. L. Bermúdez (2003) calls a *minimalist approach* to nonlinguistic thought. Taking a minimalist approach is important because the sensory-motor representations of time produced by these clocks are best described as measurements, or representations with metric structure. It would be inappropriate to characterize these representations in terms of language

[1] J. Heil (2005), p. 343.

or particularly, linguistic-propositional attitude psychology.[2] More specifically, belief-like representations about duration in animals (something like epistemic entitlements) and the representations that ground beliefs about duration in humans need not depend on linguistic capacities.

The minimalist approach, as Bermúdez describes it, is "an alternative way of construing the project of explaining the behavior of nonlinguistic creatures in psychological terms" (2003, 62). Bermúdez continues:

> The minimalist proposal is to take the psychological states attributed in such explanations to be nonpropositional, analyzing them on the model of perceptual states rather than propositional attitude states. The thoughts attributed to nonlinguistic creatures on the minimalist approach are context-bound, essentially tied to the creature's capacities for action and reaction, perceptually vehicled, and lacking the constituent structure characteristic of propositional thought (Bermúdez, 2003, 62).

The representational outputs of the clocks satisfy these properties. They are context-bound because they emulate environmentally-relevant periods or intervals. They are essentially tied to the creature's capacity for action and reaction because they are a crucial part of the sensory-motor system. In the case of the circadian clock, its representations are used to calculate the time of occurrence of periodic events, which underlies anticipatory feeding behavior. Circadian clock representations are also used to compute calculations for sun-compass navigation, which are critical for the animal's capacity to successfully interact with its environment. Similar considerations apply to the representations of the stopwatch, which, as I mentioned in the previous chapter, are critical to temporally integrate cross-modal sensorial information. These representations are also perceptually vehicled, because they rely on environmental cues, dependent upon information from the senses, as evidenced by the previous examples concerning anticipatory behavior, navigation and cross-modal sensorial information. And finally, these representations lack the compositional structure characteristic of propositional thought (See Montemayor and Balci, 2007). I will expand on representational structure in the next two sections, in which I define the sensory-motor representations of the clocks as *metrically structured analog* representations. Provisionally, it suffices to characterize these representations as

2 Even in the case of propositional attitude attribution it is debatable that a linguistic-propositional model should be adopted instead of a more minimalist one. For a detailed discussion of this issue see R. J. Matthews (2007) and references therein.

metric representations that lack the *syntactic* structure of propositional thought.[3]

The minimalist approach adequately captures some of the main characteristics of the outputs of the clocks. However, as Bermúdez acknowledges, the minimalist approach is not sufficient to give a complete account of nonlinguistic thought (2003, 62). Bermúdez focuses on two cases left out by the minimalist approach, i.e., the re-identification of *particulars* and *instrumental beliefs* in nonlinguistic creatures. But the minimalist approach *per se* is also insufficient to fully characterize the representational outputs of the clocks because it says nothing about their *structure*. It states that such structure must be non-propositional, but it does not go beyond this negative characterization.

As mentioned, I will argue that a defining characteristic of these clock representations is that they have metric structure, by which I mean that their main computational characteristics make possible the calculation of a distance function. This function encodes and preserves information about the duration of phases or intervals. In the case of the circadian clock, information about the distance between two phases is preserved in memory as time of occurrence. In addition, the phase information of the circadian clock can also be used to compute spatial information, as in the case of the ephemeris function that calculates the sun's azimuthal position. With respect to the stopwatch, distance information about intervals is preserved to compare the durations of such intervals and to anticipate non-periodic events.

One can also describe the metric structure of the outputs of the clocks by defining them as measurements of temporal information. The measuring devices would be the clocks, one of which measures temporal information periodically, in terms of phases, and the other aperiodically, in terms of intervals. In order to be interpreted as measurements, the outputs of the clocks must be somehow correlated with numbers. As Bertrand Russell (1937) says: "Measurement demands some one-one relation between the numbers and magnitudes in question—a relation which may be direct or indirect, important or trivial, according to circumstances."[4]

---

[3] Actually, even this is questionable because it is not obvious that propositional thought *itself* has syntactic structure, although clearly representations of it do. See Matthews (2007) for the case against construing thought (specifically propositional attitudes) syntactically.

[4] See B. Russell (1937), p. 176.

The clock representations would have a direct and non-trivial mapping relation with numbers, since there are no *intermediate* magnitudes or measurements involved in the mapping between periods and intervals and their representations by the clocks, which can be mathematically understood in terms of degrees of a circle or segments of a line. But the introduction of numbers seems *prima facie* unwarranted. Not all clocks need a numeric counter to work properly and, in principle, none of them requires a counter as a component to accurately measure time. It is the *user* of a clock that benefits from a counter, and the artificial introduction of numbers to analogize clock representations to measurements might make the one-one relation, mentioned by Russell, problematic.

In the spirit of the minimalist approach, it would be desirable to avoid assigning more structure to the clock representations than is required to accurately define them. Particularly, it would be desirable to avoid over-assigning the structure of the real numbers to the structure of the outputs of the clocks. Eli Dresner (2004) defines this over-assignment of structure as: "the unwarranted assumption that every numeric relation holding among two (or more) numbers represents some empirical, physical relation among the objects to which these numbers are assigned as measures (e.g., of temperature)" (2004, 467). Dresner exemplifies the over-assignment of structure in the case of temperature as follows:

> Twenty is two times ten. The length of a body of 20 centimeters is twice the length of a 10-centimeter body. But is the temperature of a body at 20 Centigrade two times the temperature of a body at 10 Centigrade? No, it is not. This is easy to see by converting the temperatures into Fahrenheit: 20 Centigrade is 68 Fahrenheit, 10 Centigrade is 50 Fahrenheit, and thus the first temperature is no longer two times the other. As these are the same two temperatures that are being measured, each time in a different scale, we conclude that there is just no fact of the matter in temperature reality of one body's temperature being x times the temperature of the other (Dresner, 2004, 467).

To think that the temperature of a room at 20 Centigrade is twice the temperature of a room at 10 Centigrade is equivalent to over-assigning the structure of these numbers to the structure of temperatures. The numbers (20 and 10) and the scale in Centigrade degrees *measure* the temperatures in question. But a different scale will yield different numbers and different relations among these numbers. There is a one-one relation between the numbers and the magnitudes in question (temperatures). But different scales produce different numbers.

The relation between numbers and magnitudes is important to define the metric structure of the outputs of the clocks, because if the relations among numbers are not direct mappings of relations among phases of periods or segments of intervals that *fully capture their structure*, then the mapping relation determined by the distance function described above will be constitutive of a *homomorphism*, rather than an *isomorphism*. As Dresner says:

> What is required for measurement is only a *homomorphism* from the empirical structure to the mathematical one (i.e., the numbers), not an *isomorphism* between the two structures. That is, the empirical structure must be mapped into the mathematical one *but not necessarily the other way around*: there could be extra structure in the abstract mathematical entity that does not reflect anything in the empirical structure being measured (Dresner, 2004, 470).

It is adequate to define the mapping between the metric structure of the outputs of the clocks and numbers as a homomorphism. The user of a clock might need numbers to measure periods or intervals, but these numbers may not reflect the structure of such periods or intervals. In other words, the counter may have extra structure that does not reflect the structure of the periods and intervals. The notion that the type of mapping relation required for measurement (and for mental representations that can be construed as measurements) is a homomorphism, receives support from the psychological literature. For example, Gallistel and King (2009) define representation in psychology in terms of homomorphism. They characterize the behaviorally relevant mapping constitutive of a representation as a "functioning homomorphism" because of its causal efficacy (2009, 70).

In general, it is true that measurements and mental representations that are analogous to measurements are best described as homomorphisms. However, in the specific case of periods and intervals, the mapping between the corresponding mathematical structure and the structure of representations of periods or intervals may adequately be understood as an isomorphism. Some philosophers, for instance Kant and Schopenhauer, have claimed that the origin of the mathematical continuum is our sense of time, or the "form of inner sense." Presumably, this "inner sense" is the source of an isomorphism between numbers and the sense of time.

In order to defend the claim that the mapping of numbers and periods or intervals may adequately be understood as an isomorphism, I will first describe in more detail the difference between a homomorphism and an isomorphism. I will then explain why the mapping between the structure of the outputs of the clocks and the corresponding mathematical structures is constitutive of an isomorphism. However, I shall emphasize that a homomorphism is *sufficient* for successful behavior and reliability. Thus, a homomorphism is sufficient to satisfy the uniformity and unity of time constraints. But obviously, since an isomorphism is a stronger form of homomorphism, it *also* satisfies these constraints.

A homomorphism is a relation among structures: it is a one-to-one mapping between the constituent objects of the represented structure to the objects of the representing structure, such that all the relations that exist among the objects of the represented structure are preserved among the objects of the representing structure. So by definition, a homomorphism is a structure-preserving mapping. In the case of temperatures and numbers, the numbers preserve the order relation that exists among temperatures, which also preserves mathematically relevant properties, such as their compliance with the axioms of addition.

An isomorphism is a type of homomorphism: it is a *bijective* homomorphism, which means that the one-to-one mapping is symmetric and preserves the structure from the representing structure to the represented structure and vice versa. In the case of the temperature example, one instance of a relation dependent upon the structure of numbers (*being twice* the magnitude) surpasses the structure of temperatures because it is not preserved in such structure, and this is the reason why such a mapping is a homomorphism, rather than an isomorphism. If two structures are isomorphic, then there is a two-way homomorphism, i.e., every structural relation among the constituents of each structure is preserved and both structures are identical.

The mapping between the phases of a cycle (the *represented* structure) and the numbers that mark the 360 degrees of a circle (the *representing* structure) is clearly a homomorphism, because all the relations among the phases are preserved by the one-to-one mapping between phases and degrees. The same is true about the segmentations of an interval and the numbers of a line. But are these mappings constitutive of an isomorphism too, i.e., are the numeric structures of the circle and the line identical to the structures of periodic cycles and intervals? In both cases, it is plausible to say that the answer is affirmative, as Kant and Schopenhauer thought.

The 360 degrees of a circle can be mapped to the phases of a periodic cycle and obtain the same structure without making under- or over-assignments of structure. If this is true, then properties of the degrees of a circle must be identical to the properties of the phases of a periodic clock. I will give two examples of such cases of identity of structure in the circadian clock. First, the numbers 0 and 360 mark the same degree of the circle, and this property is found in the circadian clock's concluding phase of a cycle, which also marks the beginning of a new cycle. Second, in the case of addition, a sum of degrees that exceeds 360 degrees gives as a result a degree smaller than 360 degrees. In the case of the phases of the circadian clock, events happening beyond the cycle of 24 hours are represented as happening *within* the cycle, which is why the circadian clock's representations confound moments in time. For instance, $180° + 360° = 180°$ (not $540°$). If a circadian clock is at phase 'noon' (half its cycle) and you isolate it for a whole cycle of 24 hours it will not shift: its next phase will be noon, which is structurally identical to the addition $180° + 360° = 180$. Additions of numbers, therefore, do not over-assign structure, unlike the case of temperature.

The experiments I mentioned in the previous chapter exemplify other isomorphic properties between the outputs of the circadian clock and the mathematical structure of the circle (e.g., representations that depend on phase information and entrainment). A consequence of the isomorphism that exists between the structure of the outputs of periodic clocks and the circle is that all periodic clocks are structurally and functionally isomorphic.[5] Incidentally, the structural and functional isomorphism of all periodic clocks is based on the transitivity of identity, because isomorphism establishes the identity of two structures.

What about interval clocks? Interval clocks and their representations are isomorphic to the mathematical structure of a line.[6] The line is understood mathematically as the set of real numbers because of its *continuity* and *ordered structure*. The real line (the set of all real numbers) is a totally ordered set because, given any two real numbers, either they are identical, or one of them is bigger than the other.[7] However, animals and humans with the capacity to represent intervals (based on the stopwatch) never represent open or infinite intervals. Thus, the isomorphism between the line and an interval is best captured by a closed interval, such as $[0, 1]$.

---

[5] I say that the clocks (periodic and interval) and their *representations* are isomorphic to mathematical structures because these mechanisms represent by *emulating* periods and intervals. Thus, what holds for the functioning structure of the clocks holds for their representations.

[6] In mathematics, topological considerations complicate the definitions of 'point,' 'line,' and 'space.' However, the *basic* definitions of 'circle' and 'line' that I am using are the ones required to capture the structure of periodic and interval clocks.

[7] This is not a claim about the structure of *physical* time, which is, according to most physicists, also continuous, but could turn out to be discrete. This claim about continuity is only meant to capture *representations* of intervals.

The number 0 (it could actually be any number within a close interval) corresponds to time *t1*, which denotes the activation of the stopwatch, and the number 1 corresponds to time *t2*, which denotes the deactivation of the stopwatch. Intermediate numbers between 0 and 1 correspond to intermediate time segments of the interval. The properties of the real numbers (continuity and order) preserve the structure of the moments of an interval and vice versa. As a consequence, all interval clocks are structurally isomorphic.

Historically, the isomorphism between intervals and the real numbers has played a very significant role in the development of mathematics. It is a central topic in the foundations of mathematics and some mathematicians, following in the footsteps of some philosophers, have characterized our capacity to represent temporal intervals as the origin of mathematical thought. For example, L. E. J. Brouwer (1907) described our capacity to perceive time intervals as the basic intuition of mathematics.[8] As I mentioned in chapter 2, one of the advantages of interval timing is that it makes possible visualizations of time that capture our intuitive notion of time. These visualizations are constantly used in mathematics.

In summary, measurements are best described in terms of homomorphism. In the specific case of periodic and interval timing, I argued that the mapping between the structure of the outputs of the clocks and their corresponding mathematical structures is isomorphic. A homomorphism suffices for the purposes of explaining how the clocks represent—in case isomorphism is considered to be too strong. But if I am right, and such mappings are isomorphic, why are time measurements unique in this respect? The answer could be that time, like space, is a *primitive* magnitude. This means that time is a fundamental magnitude that cannot be decomposed into other magnitudes. In contrast, magnitudes like temperature or rate are derived from other, more fundamental ones. S. S. Stevens explains:

> The classical view of measurement [...] is essentially the view that direct or "fundamental" measurement is possible only when the "axioms of additivity" can be shown to be isomorphic with the manipulations we perform upon objects. Only a few properties, such as length, weight, and electric resistance, are measurable in this fundamental way (Stevens, 1959, 22).[9]

---

[8] The philosophical pedigree of this idea goes back to at least the work of Immanuel Kant.

[9] The words within the quotes in Stevens' citation are references to N. R. Campbell's (1928) presentation of the classical view of measurement in physics, which is what Stevens is commenting on.

One of these fundamental magnitude-properties is time, as measured by periodic and interval clocks. Many other magnitudes derive from time and its combination with other fundamental or derived magnitudes. Time combined with number produces rate; distance divided by time is speed, etc. Stevens (1959), in his proposal for a theory of measurement that goes beyond the classical view, postulates four fundamental scales: nominal, ordinal, interval and ratio (1959, 25). Of these scales, two (interval and ratio) depend on timing, which demonstrates the primitiveness of measurements of time. In contrast, temperature and many other magnitudes are measured in different scales and depend on other, more fundamental magnitudes, such as mass and density.

In this section, I defend a minimalist approach to the nonlinguistic representational outputs of the clocks in terms of isomorphism. Periodic clocks and their representations are isomorphic to the mathematical structure of a circle. Interval clocks and their representations are isomorphic to the mathematical structure of a line. The numbers that mark the degrees of a circle and the real numbers of the line are necessary to characterize and manipulate these temporal representations. This solves the two fundamental problems of numerical measurement theory as follows.[10]

It solves the *representation problem* because it justifies the assignment of numbers to phenomena (periods and intervals) by showing that the numeric structure of the circle and the line preserve the structure of the empirical systems in question (i.e., the representational outputs of the circadian clock and the stopwatch). In addition, it solves the *uniqueness problem* by demonstrating that such mapping is isomorphic and independent of scale (unlike temperature or weight). Since the mappings are isomorphic, there is no over-assignment of structure by interpreting numeric relations as empirical facts about the clocks and their representations. In the next section, I discuss some theoretical implications concerning the metric structure of these representations, expand on the specific metric properties of periodic and interval clocks, and provide experimental evidence of the impact of these representations on the behavior of animals and humans.

[10] See P. Suppes and J. Zinnes (1967) for a clear presentation of measurement theory. See also R. J. Matthews (1994) for a measurement-theoretic account of propositional attitudes that addresses these problems.

3.2. Metric Structure

In the previous section, I defined the metric structure of periodic and interval clock representations in terms of isomorphism and a distance function. I shall now elaborate on this definition, emphasizing how it relates to the specific characteristics of the two types of clock representations. The distance-preserving function is a computation based on the two isomorphic mappings described previously, which correspond to the functional structures of each clock. I will characterize only the *outputs* of the clocks as representations, because only these outputs are used to attribute times to environmental particulars and are used at the organism level.[11]

The purpose of the previous section was to define temporal representations in terms of a nonlinguistic isomorphism that preserves distance relations, also known as an *isometric isomorphism*. This definition is meant to capture the specific characteristics of the clocks' representations and may not be adequate to define other representations, even within the sensory-motor system, because temporal representations are not decomposable into other metric representations, unlike other sensory-motor representations (e.g., speed and ratio).

The metric structure of the representations of the circadian clock and the stopwatch is framed by the specific characteristics of each clock, which were described in chapter 2. The advantages and disadvantages of each clock are manifest in the ways in which the distance function is computed. I will first discuss the metric structure of the representational outputs of the circadian clock and then describe the representations of the stopwatch. In both cases, I will rely on experimental evidence.

3.2.1. *The Metric Structure of the Outputs of the Circadian Clock*

In the previous chapter, I described experimental evidence that shows how the nervous systems of insects (bees in particular) are able to compute a solar ephemeris function using information from the circadian clock. More specifically, phase information concerning the temporal cycle of the circadian clock is interpreted spatially, providing the animal with

[11] The *inputs* of the circadian clock include signals concerning gene transcription, phototransduction, DNA to protein translation, and hormonal cycles, among many other biochemical signals. The stopwatch's inputs include cross-modal sensory signals and complex neurological network activity necessary to emulate intervals. It would be inappropriate to characterize all of these very different signals as representations.

a direction to orient itself. This temporal phase information of the circadian clock can be used for sun-compass navigation because it *continuously varies* with the angular positions of the sun as it moves across the sky. It is precisely because the circadian clock is a mechanism that updates its information continuously, that an isomorphism between the phases of the cycle of the rotation of the earth on its own axis and the phases of the cycle of the circadian clock can be established.

Evidence from experiments with ants further supports the thesis that the information computed by the clock is continuously updated. For example, the distance function used for sun-compass navigation utilizes outputs of the circadian clock, or readings of the clock that indicate its current phase, exploiting the isomorphism between the cycle of the circadian clock and the cycle of the rotation of the earth on its own axis. These readings appear as a variable in the function, which changes value according to the phase of the clock. In other words, the metric structure of the outputs of the circadian clock makes possible the computation of *other* metric representations, such as the calculation of the azimuthal position of the sun, because it provides a temporally framed *representation space* of possibilities based on phase-related information.[12] I proceed to explain the experimental evidence in more detail.

In the case of the desert ant, *Cataglyphis fortis*, two hypotheses were tested to determine how it computes the solar ephemeris function. The extrapolation hypothesis postulates that the ant uses the most recent encoding of the position of the sun and then extrapolates its current position. In contrast, the interpolation hypothesis states that the ant computes the ephemeris function by linearly interpolating memorized positions of the sun by *filling the gaps* of time when the ant had no environmental cues, using circadian clock information. Experiments in which ants were trained during the morning but tested at night with moonlight or artificial light confirm that they are *interpolating* the position of the sun based on information from their circadian clock (See R. Wehner and R. Müller, 1993). These results have also been confirmed in the honeybee, as mentioned in the previous chapter. Moreover, as Wehner and Müller showed:

> If ants are restricted, from the very beginning of their outdoor activities, to forage only in the early morning hours and are later tested for the first time

[12] See R. J. Matthews (1994) for an account of how representation spaces structured by metric mappings provide satisfaction conditions that allow for semantic evaluability and inferential relations.

> in the late afternoon, they expect the solar azimuth to have moved through about 180°.

Wehner and Müller interpret this result as evidence that "*Cataglyphis* is informed innately about one general spatiotemporal aspect of the sun's 24-h course, namely, that the angular positions of the solar meridian at sunrise and sunset lie opposite to each other" (1993, 333). This innate knowledge is easy to explain by circadian interpolation, but not by inferential extrapolation. In any case, this finding certainly confirms the automatic, self-sufficient nature of the circadian clock. Because of entrainment, the outputs of the circadian clock can be used for sun-compass navigation. Outputs are constantly updated and preserve an isometric isomorphism with respect to the cycle of the rotation of the earth on its own axis in a very reliable manner. They thereby provide a reliable representation space of possibilities, based on phases of the clock, which can be interpreted in terms of location (i.e., the position of the sun).

The interpolation model assumes that the times of occurrence of certain cues that correlate with phases of the circadian clock are somehow encoded and used to calculate the solar ephemeris function. Otherwise, the ants could not compute this function, because they would have no phase information in memory to interpolate. As I explained in the previous chapter, memories for time of occurrence are very important circadian clock representations. In this subsection, I will briefly describe how these memories are registered through a process that is entirely dependent upon the metric structure of the circadian clock's representations.

Because oscillators have circular trajectories in a phase plane, they can be described by an angle, or phase-angle. A very important characteristic of periodic clocks, discussed in the previous chapter, is that they are always at a phase of their cycle. Based on the isomorphism between the cycle of the circadian clock and the structure of the circle, phases can be analogized with angles, and computations of angles can encode information about phases. If one represents such computations in a Cartesian coordinate system, one can graphically show that animals encode the time of occurrence of events based on readings of the circadian clock in terms of the sine and cosine of the phase. As Gallistel explains:

> The sine-cosine representation of angle as a function of time plots the values of the state variables as functions of time. When the maximum and minimum values of the variables are set equal to one, then these functions are the sine (y-variable) and cosine (x-variable) functions. By recording the momentary values of these variables, a system specifies a momentary state of the oscillator (a reading of the clock). This yields a specification of time

> unique up to a translation by an integer number of periods along the temporal axis (Gallistel, 1990, 233).

These mathematical computations are possible because of the metric structure of the circadian clock's representations, and they are the basis for the memory for time model I described in the previous chapter.[13]

However, the *disadvantages* of periodic clocks affect the scope of these computations because the recording of moments in time in terms of sine and cosine can only distinguish moments within a period of 24 hours. As is the case with any other periodic clock, the circadian clock's representations confound moments in time that go beyond its cycle. It is possible, and actually quite probable, that some animals have the capacity to phase-relate *infradian* cycles to the circadian clock's cycle, which would allow them to distinguish moments beyond the period of 24 hours. This may depend on the lifespan of an animal. Animals with a very short lifespan may do very well without representations of time that go beyond the 24 hours cycle of the circadian clock. Other animals, like birds, may distinguish events that do not happen daily, but monthly (See Gallistel 1990, 235).

Another important consideration, besides lifespan, is brainpower: the more complex the nervous system of an animal, the more capacity for storing and manipulating outputs from the circadian clock. If an animal has a *semi-hybrid* clock that registers circadian clock times analogously to a calendar, then representing periods of time that are much longer than 24 hours would be a relatively easy task.[14] This seems to be the case with some birds from the corvid family, particularly scrub jays.

In an impressive series of experiments, Clayton, Dickinson, and their collaborators (1998, 2003a, 2003b, 2006) have shown that scrub jays not only have a semi-hybrid clock that allows them to organize temporal information in a calendar-like fashion, but that they also use such information to create a spatiotemporal representation of their caches that includes their location, rate of decomposition, and even whether or not other jays were looking when they made the cache (See Gallistel, 2008; and C. Montemayor, 2010, for a review of these findings). The amazing

---

[13] If these memories of times are used to calculate an *interval* of time or if an interval clock is used to measure the durations between phases of the periodic clock, there is no *genuine combination of clocks*, as the periodic segmentation principle states (see chapter 2). The periodic clock's phases are not *determined* by intervals. Rather, the interval clocks are *used to measure* the duration of segments of the periodic clock's cycle.

[14] For discussion of hybrid clocks and the principles governing them, see chapter 2.

episodic-like memory capacity of the scrub jay allows it to represent complex cognitive maps for caches. Although scrub jays make more than *thirty thousand* different caches during their life, doing so over vast areas of landscape, they successfully retrieve these caches during the days and months where food is scarce (See also S. B. Vander Wall, 1990).

Regardless of how one characterizes the types of memories manifest in the scrub jay's behavior, it is clear, at a minimum, that this evidence suggests the existence of a very sophisticated type of memory.[15] The outputs of the clocks need not have symbolic structure and actually, in section 3.3., I will argue that the metric structure of the outputs of the clocks is in an analog, rather than digital, format. However, readings of the clocks by other regions of the brain, particularly regions that store information symbolically, will *interpret* and *store* the information from the clocks in a different format, thus creating a complex semi-hybrid clock.

Notice that even when distance functions that are not exclusively temporal (such as the ephemeris function) are computed by other symbolically driven regions of the brain, the metric structure of the outputs of the clocks, in this case the circadian clock, is what makes these computations *distance-preserving* ones, because they are isomorphic to the structure of environmentally relevant periods.

3.2.2. *The Metric Structure of the Outputs of the Stopwatch*

In chapter 2, I showed that the scalar property of interval timing corresponds to the scale/resolution tradeoff characteristic of interval clocks. As a consequence, the distance function constitutive of the metric structure of the outputs of the stopwatch manifests this tradeoff as a degree of error. I offered evidence that demonstrates how the stopwatch's degree of error in calculating intervals increases proportionally with the duration of the interval. In this section, I shall briefly describe how this metric property relies on the isometric isomorphism that exists between the outputs of the stopwatch and the intervals that they emulate.

As I mentioned in chapter 2, Gibbon (1977) presented the scalar timing theory in order to explain the scalar property of the stopwatch's interval timing. Evidence in support of scalar timing shows that there is a latency

---

[15] Gallistel claims that these findings demonstrate the existence of a symbolic read-write memory, similar to the memory of a computer. (See Gallistel and King, 2009). I will remain neutral about this issue and only acknowledge that the evidence indeed demonstrates a highly sophisticated kind of memory in animals and humans.

difference between the experienced intervals, in which food (or another reward) is randomly given to an animal, and the intervals between peaks of activity, which indicate that the animal expects food. The important discovery concerning the latency between, as Gibbon named them, the objective (external) and the subjective (internal) intervals, is that it is governed by a scalar factor. Gallistel describes one of the relevant findings, concerning a procedure with fixed intervals experienced by rats, as follows:

> The rat's peak response rate occurred at 24 seconds where there was a 20-second reward latency and at 48 seconds when there was a 40-second reward latency. In both cases, the peak, which is taken to indicate when the rat expects the food, occurs at an elapsed interval equal to the correct interval multiplied by 1.2. It is this finding (and numerous similar ones in other tasks) that leads to the term *scalar timing theory*. [...] The model developed by Gibbon and his collaborators postulates that the remembered duration of an elapsed interval is its experienced duration multiplied by a scalar factor, which varies from animal to animal (Gallistel, 1990, 301).

The value of the scalar, or multiplicative factor, may change from animal to animal, but it is always constant within an animal. Otherwise, the stopwatch's memory component would not be constitutive of a *reliable* clock.

To illustrate this, imagine the following (somewhat idealized) scenario. Suppose that three people have to measure intervals, say the durations of different songs. They each have a reliable sand clock, with a constant rate and numerically marked intervals, but each clock contains different sand materials, say carbon, granite, and salt. Every time they stop their clocks and measure an interval, they empty the content and dehydrate the sand material, compressing it into a small brick that they put into a container, which classifies songs according to their numerical order. Suppose that someone asks them, immediately after they compressed one of their measurements, to report the duration of that interval. They would add water and the sand would grow back to its original size, but given that the materials have different density, they might not grow back to their *exact* original size. However, as long as they always add the same amount of water, the sand would grow at a constant *ratio*. Thus, one of them would approximate the interval, say 1 minute, very closely: 1.01 minutes (a factor of 1.01). The other two measurements might not be as accurate, but they would not be too far off target: 1.1 and 1.2 (factors of 1.1 and 1.2). Had they measured an interval of 2 minutes, they would have reported 2.02, 2.2 and 2.4 respectively. These numerical values differ, but they preserve the length of the original interval in virtue of a scalar factor.

The *scale/resolution tradeoff* characteristic of all interval clocks is manifest in this metric scalar. If the interval is short, the resolution is good. But as intervals get larger, the resolution drops. If an interval has a duration of 20 minutes, the corresponding values of the three previous clocks' information-retrieving mechanisms will be: 20.2, 22, and 24. For an interval of 100 minutes, the values would be 101, 110, and 120. 20 minutes seems to be a very large difference, compared to the 12 seconds of the first example (.2 minutes). However, the *degree* of accuracy or reliability is the same: 1.2, even though the resolution of the clock drops as the duration of the interval grows.

The scalar property of interval timing illustrates a very important characteristic of the outputs of the clocks, besides their metric structure, namely that they are *approximate representations*. In the next section, I explain that this property is crucial to understand why the outputs of the clocks are *analog* representations. I shall now describe in more detail why the scalar property of interval timing demonstrates that the isomorphism between the representing interval (which explicitly emulates an interval by an accumulation or decay process) and the represented interval, is *distance preserving*.

The scalar factor governing the outputs of the stopwatch is also known as Weber's law. Weber's law also governs other kinds of magnitude-based representation, such as number and ratio. Since time is a primitive magnitude (it cannot be decomposed into other, more primitive magnitudes), preserving temporal metric relations is fundamental for computing other non-primitive magnitudes. The case of the computation of the ephemeris function is one example, but computations of rate and speed are other, equally relevant examples.

Weber's law is a ubiquitous feature of the *comparison* of magnitude representations in animals and humans.[16] It captures the scalar factor of interval timing, and it is expressed by the formula $\Delta I/I = k$. In words, the formula says that the difference threshold (ΔI)—the minimal change required for *discrimination*—, divided by the value of the initial stimulus or magnitude, is constant (k). The value of (k)—the constant—has to be

[16] Conformity to Weber's law has been reported in adult humans (Cordes, Gelman, Gallistel, and Whalen, 2001; Moyer and Landeuer, 1967; Whalen, Gallistel, and Gelman, 1999), infants (Xu and Spelke, 2000), and animals (Cheng and Roberts, 1991; Church and Gibbon, 1982; Gibbon, 1977) both in the temporal and number domains. Conformity to Weber's law is considered to be a signature of magnitude representations (See Gallistel and Gelman, 1992, and references therein for review).

found through experiment because, as mentioned, this value, which is the scalar or multiplicative factor, varies from animal to animal.

Changes in value are not noticeable within the difference threshold. This is why Weber's law is frequently explained in terms of "just noticeable differences." In the previous case of the interval with a duration of 100 minutes, the values were 101, 110, and 120. Weber's law determines that in the first case, the constant is 101–100, which is 1 (the difference threshold or ΔI). This value (i.e., 1) divided by 100 is .01. The same formula applied to the other values results in .1 and .2, which are indeed the differences in error that are kept constant, plus the unit interval. This means that the multiplicative factors that are kept constant for these values are the original magnitudes' values, represented by 1, *plus* the scalar factor responsible for the difference threshold, which gives the correct values: 1.01, 1.1, and 1.2.

Weber's law explicitly captures the approximate nature of magnitude representations. For example, in the case where the interval is 100 minutes and the scalar factor, or degree of accuracy is 1.2, the represented or retrieved interval will be 120 minutes. This means that the difference threshold is 20 minutes. Represented intervals that are within this threshold, i.e., experienced intervals whose values lie between 101 and 119 minutes, will be considered as *equal in duration* by the creature whose scalar factor is 1.2, i.e., 100 minutes. It is not possible for such creature to discriminate intervals that lie within this threshold.

As mentioned, Weber's law has also been confirmed in the number and rate domains in animals and humans. It applies to the noticeable differences in intensity of stimuli and differences in weight and sound, and it also seems to apply across the board with respect to magnitude-based representations. This shows that the factor responsible for the scalar variability of interval timing is not the result of an over-assignment of structure based on the numeric values of a *scale* to the metric structure of the outputs of the clock. Rather, it is one of the most critical *psychophysical laws* that govern the retrieval of information concerning isomorphic (in the case of time) or homomorphic mappings between stimuli and representations.

The metric structure of the outputs of the stopwatch makes possible the computation of distance-preserving mathematical calculations, like those based on the outputs of the circadian clock. Through different experimental procedures, researchers have found that intervals can be *added* and *subtracted*, particularly in the so called "time-left" paradigm, in which an animal has to choose one of two options that will provide a payoff in a specific amount of time. Animals have to compute ratios between intervals in memory and experienced intervals in order to choose optimally. Many experiments confirm that animals are capable of choosing optimally in the time-left paradigm based on the addition and subtraction of intervals (See Gallistel, 1990, 315).

These temporal ratios, which are governed by Weber's law, are based on the metric structure of the outputs of the stopwatch, and they constitute the cornerstone of interval timing. These metrically structured representations make possible the formation of representation spaces that determine options for animals. They specify which course of action is the most optimal one, based on the lengths of the intervals they compare. For example, in the time-left paradigm, the animal uses the metric information provided by the outputs of the stopwatch, plus information concerning rate stored in memory, to decide which of two options (pushing lever A or B) will produce a comparatively better reward, and animals are incredibly accurate in performing this time-sensitive task. The evidence in support of these metric features of interval timing is abundant.[17]

At this point, it is important to briefly reflect on the richness of the metrically-structured representation spaces of possibilities for action and motor control that the outputs of the clocks reliably produce. The nervous system uses constraints on the uniformity of time and the units of time in order to frame motor control, thereby structuring action with reliable measurements of time. This allows creatures with nervous systems to store memories of successfully-timed behavior, compare reliably-measured intervals, and predict future events. A vast horizon of possibilities opens up for creatures that have these capacities. For instance, forms of counterfactual reasoning are possible: "If I press lever A, then I will get less food in a briefer amount of time, but if I press lever B, I will get more food in just a slightly longer amount of time. So I should press lever B and wait a bit longer." The reliability of the clocks is what makes these

---

[17] I mentioned in the previous chapter that there is an alternative model for interval timing that dispenses with the stopwatch, favored by Gallistel and King (2009). They base their objection to the stopwatch mainly on a theoretical difficulty they call "the problem of the first interval." They postulate a different model based on a read-write memory (a not entirely uncontroversial assumption) for interval timing dependent on oscillators that works in terms of periodic segmentation. I will not assess the merits of this alternative proposal because it assumes the metric structure of interval and periodic *timing*, and is thus compatible with everything I say in this chapter. However, I shall assume that the stopwatch is the mechanism responsible for interval timing because it is the most intuitive model, as well as the most popular account among psychologists. Moreover, there are no differences between Gallistel and King's alternative model and the stopwatch with respect to experimental predictions.

complicated decision-making processes feasible computations for the nervous systems of animals and humans.

To conclude, the outputs of the circadian clock and the stopwatch are metrically structured. There are two independent systems for the representation of time, one for periods and the other for intervals, which have been confirmed to have metric properties in a vast number of experiments. However, it is important to determine how these metric-temporal representations fit into a broader spectrum of mental representations. Some philosophers have categorized mental representations in terms of analog and digital computational formats. This distinction, which originated in engineering, is very useful for categorizing the outputs of the clocks. In the next section, I explain why the outputs of the clocks are *analog representations*, and I provide an account of how they might interface with digital representations.

3.3. Analog Clock Representations

The circadian clock generates phase information that the brain uses for registering the time of occurrence of environmental particulars and, among other metric functions, the solar ephemeris function. The stopwatch generates information concerning the duration of intervals (independently of phase information), which is also metrically structured. However, establishing the metric nature of these representations is not sufficient to determine *what kind* of representations the outputs of the clocks are. Of all the categories for mental representation that have been put forward in the philosophical literature, the distinction between analog and digital formats of representation is the most useful to classify the metric outputs of the clocks.[18]

I will rely upon five criteria to characterize analog representations and argue that the outputs of the clocks satisfy all of the requirements imposed by these criteria. Of these criteria, only the first three apply distinctively to analog representations. The other two apply to representations in general, but they are of particular relevance for analog representations. The first criterion is that there is always loss of information in any analog to digital conversion, which means that analog representations must contain more

[18] Some philosophers (e.g., J. Fodor, 2008; and F. Dretske, 1981) have argued that there may be a connection between analog and digital *formats*, and the conceptual non-conceptual *content* distinction. I shall focus on the characterization of the outputs of the clocks, and remain neutral with respect to this issue.

information than digital ones. The second is that analog representations must be continuous or dense, and the third is that they must represent by approximation. Finally, the fourth and fifth criteria are that *any* representation must allow for *misrepresentation* and also for *cognitive integration*.

Before proceeding, I shall introduce an important caveat. It is easy to confound properties of the physical instantiation of a computational process with the *format* in which such a process is computed. For example, it is frequently assumed that analog computation depends on continuous physical processes and that digital computation depends on discrete ones. This assumption originates from the fact that analog computation manipulates signals and frequencies, physically instantiated by waves and charges that continuously vary their value. In contrast, digital systems of computation manipulate symbols with a particular meaning or value assignment. When numbers are used to characterize analog processes, these processes are frequently represented by a continuous interval, say the real numbers that lie between 0 *and* 1, or [0, 1]. In contrast, digital processes are assigned the binary values 0 *or* 1. But notice that these number assignments are not characterizing the *media* that instantiates a computational process. Rather, they are characterizing *how the information in the computational process is formatted* or how the user is *manipulating* the information.

In other words, the continuity of the real numbers is used to represent the continuous computational states of an analog computer and the discreteness of a string of numbers is used to represent the discrete computational steps of a digital computer.[19] Confusion stems from assuming that these properties (continuity and discreteness) are also properties of the media that instantiate the states of an analog or digital computer.

In fact, some philosophers have argued (e.g., J. Haugeland, 1998; and D. Lewis, 1971) that the distinction between analog and digital is problematic if applied to the media that instantiate computational processes. To illustrate this point, one can count discrete numbers, and perform arithmetic operations on integers by using discrete amounts of electric charges (a continuous medium). One can also perform approximate measurements of time by using a clock that instead of water uses pebbles

19 Interestingly, the continuity or discreteness of computational processes depends a lot on how time is represented. See C. Moore (1990, 1996) and Mycka and Costa (2007) and references therein for discussion and potential applications of analog computations with a continuously represented time.

(a discrete medium).[20] The first computation is digital because each discrete amount of electricity is used as a symbolic stroke in a tally notion system, and the second computation is analog because no individual pebble is used as a symbol. Thus, the distinction between analog and digital concerns formats of representation, not media.

This clarification is very important because otherwise the notion of analog representation is doomed to be imprecise. In order to provide a precise definition of analog representation, one must specify criteria that representations must satisfy in order to be characterized as analog. As Zenon W. Pylyshyn (2007) explains,

> Despite the existence of clear and easily understood cases of analog processes, and despite the frequent references made to this notion, it remains poorly understood. In particular, it has turned out to be extremely difficult to give an acceptable set of conditions for something being an analog (Pylyshyn, 2007, 163).

Pylyshyn highlights the polysemy of the term 'analog.' For instance, Nelson Goodman (1976) and Lewis (1971) used it to characterize processes or representations that are continuous. Fodor (2007), by contrast, has argued that analog representations have no canonical decompositions into semantically interpreted constituents, without focusing on the continuity of processes or representations. I submit that this problem of polysemy originates at least partly from thinking that both medium (or vehicle) *and* representation (or format) must each be continuous.

Having made this caveat and explained its importance, I shall now discuss the five criteria that define analog representations and show that all of these criteria are satisfied by the outputs of the circadian clock and the stopwatch. The main purpose of discussing these criteria is to give an accurate characterization of the outputs of the clocks. By contrasting these criteria with the defining criteria for digital or symbolic representation and by demonstrating how temporal magnitudes satisfy these criteria, however, this section will also meet a secondary, but theoretically relevant goal, namely to provide a set of conditions that sufficiently capture analog representations in general.

3.3.1. *First Criterion: Loss of Information*

Fred Dretske (1981) has offered one of the most influential characterizations of analog and digital formats of representation. His account of

---

20 This example is partly based on D. Lewis (1971).

analog representation is based upon the loss of information that happens every time analog information is converted into digital information. This loss of information comports with the way in which digital devices perform *readings* on analog signals that contain more information than any particular digital computation or reading. As Dretske says,

> To describe a process in which a piece of information is converted from analog to digital form is to describe a process that necessarily involves the *loss* of information. Information is lost because we pass from a structure [...] of greater informational content to one of lesser information content (Dretske, 1981, 141).

Dretske illustrates this *loss of information* principle with the way in which we linguistically communicate information about the content of pictures. The content of the sentence 'the cup has coffee in it' is encoded in a picture of a cup with coffee in it. However, the picture contains much more information than the fact that there is coffee in a cup, e.g., *how dark* the coffee is, *how much* coffee is in the cup, what *color* and *shape* the cup has, etc. The propositional content of the sentence just mentioned does not capture information about these features. As Dretske says: "To say that a picture is worth a thousand words is merely to acknowledge that, for most pictures at least, the sentence needed to express all the information contained in the picture would have to be very complex indeed" (1981, 138). Dretske proposes the following definitions in order to capture the loss of information from analog to digital:

> A signal (structure, event, state) carries the information that *s* is *F* in *digital* form if and only if the signal carries no additional information about *s*, no information that is not already nested in *s*'s being *F*. If the signal *does* carry additional information about *s*, information that is not nested in *s*'s being *F*, then I shall say that the signal carries this information in analog form. When a signal carries the information that *s* is *F* in analog form, the signal always carries more specific, more determinate, information about *s* than that it is *F* (Dretske, 1981, 136).

Notice that Dretske's definitions are neutral with respect to the kind of medium that instantiates a cognitive process and focus on the form in which information is conveyed. This neutrality avoids the potential confusion mentioned before concerning media and formats. A consequence of Dretske's definitions is that analog information will always carry some kind of digital information, at least potentially. Decoding digital information from an analog signal will require a specific informational *filter*.

In the case of the circadian clock, the *readings* performed on its outputs, which are used to compute the solar ephemeris function and the time of occurrence of an event, have less *temporal* information than such outputs. For instance, both readings lack temporal information concerning periodicity, phase-fixed relationships with ultradian and infradian cycles, the degree of completion of the circadian cycle, etc. All this information is explicitly represented by the outputs of the circadian clock. Similarly, in the case of the stopwatch, information stored in memory concerning the *comparison* of short intervals may lack information concerning the specific length of the initial intervals and the rate at which the stopwatch was operating (information explicitly represented by the outputs of the stopwatch). The criterion of loss of information suits well the outputs of the clocks: temporal information is lost when it is converted from the analog outputs of the clocks to other, more specific and digital computations in the brain.[21]

However, before proceeding with the other criteria for analog representation, it is useful to make a distinction proposed by R. Cummins, et al. (2001). The question that motivates this distinction is: how should one categorize computational processes that systematically *recover* information that was lost in the analog to digital conversion? Cummins, et al. (2001) distinguish between structural encodings, pure encodings, and structural representations. Structural representations are isomorphic to the structure that they represent—as is the case with the circadian clock and the stopwatch's outputs (they are isomorphic to the structure of the period of the rotation of the earth on its own axis and to the structure of the brief environmental intervals emulated by the stopwatch).[22] Structural encodings are neither isomorphic nor homomorphic to what they represent, but one can recover the structure of what is represented through some computational process.[23] Finally, a pure encoding is a symbolic representation that is neither isomorphic nor homomorphic to what it represents, and from which it is impossible to recover the structure of what is represented.

[21] Computer models that describe how the brain can rapidly update complex information in a stable and reliable way suggest that the anatomy of the brain allows for a hybrid, analog, and digital system of computation. For example, Randall C. O'Reilly (2006, 91) argues that the prefrontal cortex has a discrete, digital character, while the rest of the cortex operates in an analog fashion.

[22] I described the mathematical aspects of these isomorphic representations in section 3.1. of this chapter.

[23] The distinction between structural representations and encodings was originally introduced by R. Cummins (1996).

Dretske's criterion of loss of information from analog to digital conversion holds for both types of *encoding*, as defined by Cummins, et al. In the case of pure encodings, the loss of information is quite significant: all information concerning structure (e.g., metric or inferential) is lost and cannot be recovered. In the case of structural encodings, some information might be recovered, but it would be extremely inefficient (for the brain or any other computational system) to recover *all* the information, because it would have been more economic not to convert it into digital format in the first place. However, such unparsimonious algorithms may be necessary in some cases (the result would be that the fully recovered structure is a structural representation, not a structural encoding).[24]

3.3.2. *Second Criterion: Continuity and Density*

Historically, in the development of computer science, the criterion of continuity and density was the most important one to distinguish analog from digital formats. The basis of this distinction is that computations that are analog vary their value assignments continuously, in a pattern that can be modeled as a continuous interval of values [0, 1]. In contrast, digital computations have either fixed value assignments, or a very limited set of possible value assignments (e.g., either 0 or 1) and need to be modeled discretely, as a set or string of symbols.

Continuity and density have also been relevant for philosophical accounts of analog representation. For example, Nelson Goodman (1976) characterizes analog representations as syntactically and semantically dense. He explains that a representation is syntactically dense "if it provides for infinitely many characters so ordered that between each two there is a third" (1976, 136). Semantic density is a consequence of syntactic density, and it consists in the impossibility of assigning differentiated and unambiguous values to discrete symbols.

As Goodman notices, density does not entail continuity, because although the rational numbers are dense, according to the definition just given, there are "gaps" between them. This is the reason why discreteness is a necessary, but not sufficient, condition for digital representation.

[24] James Blachowicz (1997) says that analog representation is constrained by resemblance. The type of isomorphism required for the *model approach* to analog representation that he defends comports with the isomorphism that exists between the outputs of the clocks and what they represent, and it is also compatible with the notion of structural representation promoted by Cummins et al.

Goodman argues that what is required for digital representation is *differentiation*, not mere discreteness.[25] Thus, according to Goodman, there can be analog representations that are not continuous. However, in the case of the outputs of the clocks, I will argue that these representations are dense *and* continuous, i.e., without gaps. Consider the following case of analog representation, offered by Dretske:

> The speedometer on an automobile constitutes an analog encoding of information about the vehicle's speed because different speeds are represented by different positions of the pointer. The position of the pointer is (more or less) continuously variable, and each of its different positions represents a different value for the quantity being represented (Dretske, 1981, 135).

Indeed, this representation is continuous. If the pointer of the speedometer feeds its information into a digital register at a particular time, the digital numeric symbol for the speed of the automobile *at that time* would be displayed. But the speedometer's pointer represents the automobile's speed *continuously*, as it unfolds in time. Speed is a continuous magnitude that is isomorphic to the real numbers because there are no gaps between speeds.[26] The speedometer represents the continuity of speed by means of the smooth transition of its pointer across the continuous surface, in which a range of speeds is represented. As mentioned previously, this is not a matter of whether the medium is physically continuous. Rather, it is a matter of how speed is being represented.

Analogously, the circadian clock continuously (without gaps) varies its phases in order to represent and emulate the period in which the earth rotates on its own axis. Each phase of the circadian cycle is like a position of the continuous surface of the speedometer, and readings of the circadian clock are equivalent to readings of the pointer of a speedometer. Similar considerations apply to the stopwatch: each segment of an interval emulated by the stopwatch is like a position of the speedometer and

---

[25] Differentiation, in computational terms, requires individuation conditions for *symbols*. These individuation conditions, in turn, demand the existence of a program that individuates symbols according to their computational role. The lack of individuation conditions for symbols in analog representation entails the lack of symbolic constituents of more complex, syntactically structured representations, which also entails the lack of canonical compositions and decompositions. This comports with Fodor's (2007) characterization of analog representation.

[26] The isomorphism between speed, time and the real numbers was a major factor in the development of the calculus, created to a large extent for the purpose of calculating instantaneous velocities on the basis of the continuity of these magnitudes.

there are no gaps between segments of intervals. The isomorphism between the outputs of the clocks and what they represent is a systematic and continuous mapping that, as explained in section 3.2., preserves the metric structure of periods and intervals. I shall now proceed with the third criterion that analog representations must satisfy: their approximate nature.

3.3.3. *Third Criterion: Approximate Representation*

The third criterion that analog representations must satisfy is a consequence of the previous criteria, because the richness of information and the lack of symbolically differentiated value assignments characteristic of analog representations imply that analog representations *represent by approximation*. The outputs of the clocks approximately represent periods and intervals by emulating them. As I explained in chapter 2, emulating periods and intervals depends on several factors. The variability of these factors may jeopardize the degree of precision and reliability with which periods and intervals are approximately emulated, e.g., *phase drifts* in the case of the circadian clock and *variations in rate* in the case of the stopwatch.

John Haugeland (1981) distinguishes analog from digital representations in terms of approximation. He says that digital devices (or systems of representation) are defined by the following four features: a) a set of types; b) a set of feasible procedures for writing and reading tokens of those types; c) a specification of suitable operation conditions, such that, d) under those conditions, the procedures for the write-read cycle are positive and reliable (1981, 216). Haugeland says that a procedure is *positive* just in case it can succeed absolutely and without qualification, and *reliable* just in case, under suitable conditions, it succeeds virtually every time (Haugeland, 1981, 215).

In contrast, Haugeland argues, analog devices (or systems of representation) are defined by approximate read-write cycles, or approximation procedures that occur within certain margins of error, such that: a) the smaller the margin of error, the harder it is to stay within it; b) available procedures can *reliably* stay within small margins of error and c) there is no limit as to how small the margin of error can be, but it will never be zero: perfect or *positive* procedures are not possible (Haugeland, 1981, 221).

If one interprets the read-write cycles in terms of the *isomorphic mapping* from the period of the rotation of the earth on its own axis to the cycle of the circadian clock and from environmental intervals to the emulated intervals of the stopwatch, then Haugeland's definition of analog representational devices accurately captures the essential characteristics of these clocks and their representations. For example, the circadian clock can reliably, under suitable conditions, emulate the period of the rotation of the earth on its own axis within small margins of error (property 'b' of Haugeland's definition of analog devices). However, if the standard for the margin of error is extremely small, (within the milliseconds range) then the accuracy of the circadian clock will drop significantly (property 'a').

Suppose that some animals evolved incredibly accurate circadian clocks. No matter how accurate these clocks are, their margin of error will never be zero (property 'c'), because they ultimately depend upon variations in environmental conditions and latitude. Adjustments to these variations require *approximation* procedures, as has been consistently demonstrated by phase drift experiments, in which an animal's circadian clock takes days to "catch-up" with the new daily cycle. In more colloquial and experiential terms, one does not eliminate the experience of being jet-lagged after traveling across several time zones by computing a symbolic function. Rather, the circadian clock has to accelerate its pace and approximate the new daily cycle by shifting its phases.

Likewise, the stopwatch's outputs seem to be tailored to comply with Haugeland's criteria. Weber's law guarantees that the margin of error will never be zero, although it will be constant in proportion to the interval, which (in addition to the stopwatch's constant rate) guarantees the stopwatch's reliability. Moreover, it is a *fundamental characteristic* of interval clocks that their accuracy decreases as the magnitude of the interval increases, which further demonstrates the approximate nature of the outputs of the stopwatch.

From a purely theoretical perspective, the first feature of analog representations mentioned by Haugeland assumes continuity and density: the smaller the margin of error, the harder it is to stay within it. If the margin of error is determined in minutes, one can make it smaller and determine it in seconds, milliseconds, or even picoseconds. In principle, one can go beyond the picoseconds range because of the continuity of temporal magnitudes, which other magnitudes share. As mentioned, the outputs of the clocks are continuous and dense. A feature based on *ratio* (at least in the case of time), such as 'the smaller the margin of error, the harder to stay within it' assumes continuity and density because the process of reducing the margin of error is in principle unbound.

3.3.4. *Fourth Criterion: Analog Misrepresentation*

Bermúdez (1995) argues that there are four criteria that any physical state must satisfy in order for it to be properly described as a *representational* state. These criteria for representational states are: a) they should serve to explain behavior in situations when the connections between sensory input and behavioral output cannot be plotted in law-like manner; b) they should admit of cognitive integration; c) they should be compositionally structured in such a way that their elements can be constituents of other representational states and d) they should permit the possibility of misrepresentation (1995, 350).

Of these four criteria, only b) and d) are relevant for the characterization of the outputs of the clocks. I will not discuss criterion a) because defining what 'law-like' means would be distracting and unproductive for the present purposes.[27] However, it should be noted that the representation spaces for possible action produced by the outputs of the clocks allow for forms of counterfactual reasoning, which produce behavior that cannot be explained in terms of *mere* causation (this is what Bermúdez must have in mind). Moreover, these representations are *used* by animals and humans at the organism level, and cannot be reduced to sub-personal or causal-like manipulations of information. So regardless of how one defines law-like relations, these representations seem to satisfy this criterion.

Criterion c) is inadequate for the present purposes because it is too restrictive: it assumes that representations must be syntactically structured by atomic symbolic constituents, which does not capture the main characteristics of the outputs of the clocks i.e., their metric structure and analog nature. I will discuss criterion d) in this subsection, and then criterion b) in subsection 3.3.5.

Criterion d), which says that representational states should allow for the possibility of misrepresentation, is not specific to analog representations, and seems to be a necessary condition for any type of representation. It is particularly problematic for analog representations to satisfy this criterion, specifically in the case of the outputs of the clocks, which are defined in terms of isomorphism. By preserving metric structure through isomorphism, the outputs of the clocks reliably emulate periods

[27] This is a controversial issue in the philosophy of science. See, for example, R. Giere (1999) and B. van Fraassen (1989).

and intervals. However, by doing so, they also significantly reduce the possibility of misrepresenting such periods and intervals. An example discussed by Bermúdez, which actually concerns a periodic clock (tree rings) nicely illustrates this point:

> It would seem that a correctness condition can be provided, such as "the rings on the tree correctly indicate the age of the tree if, and only if, the number of rings = the number of years the tree has been in existence." But would these be *genuine* correctness conditions? Many would think not. They might argue as follows. No state could count as a representational state unless it was possible for it to *misrepresent* the environment. But it is the law-like connection between, for example, the number of rings and the number of years that makes it plausible to speak of the former carrying information about the latter, and what makes it a law-like connection is the fact that the number of rings and the number of years invariably coincide. Such invariable coincidence, however, clearly rules out the possibility of misrepresentation (Bermúdez, 1995, 344–345).

If one replaces 'law-like connection' with 'isomorphism' one can similarly argue that the possibility of misrepresentation in the case of the outputs of the clocks is ruled out. If tree rings *invariably coincide* with numbers of years, then how could they possibly misrepresent information about years? This seems to be a particularly pressing objection against the isomorphic representations that constitute the outputs of the clocks, because if they invariably coincide with metric features of periods and intervals, then how could they possibly misrepresent such periods and intervals?

In chapter 2, I argued that tree rings are registers of a periodic clock, i.e., the rotation of the earth around the sun. But these are registers of information that humans have to interpret, and interpretations can clearly misrepresent. The tree registers environmental changes that invariably coincide with the period of the rotation of the earth around the sun. But the tree has no access to this information. In contrast, animals with hybrid-clocks, like the scrub jay, that register time of occurrence in their brain, have access to temporal information stored in their memory. For these reasons, it is problematic to call tree rings 'representations.'

Nonetheless, Bermúdez is right in claiming that a representational state must allow for the possibility of misrepresentation. Is there a way in which the outputs of the clocks could, in spite of their isomorphism with periods and intervals, allow for misrepresentation? The answer is that there are indeed ways in which the outputs of the clocks may misrepresent. Explaining them requires a characterization of the clocks as causally-driven mechanisms of information processing, analogous to other perceptual processes. My argument relies on characterizations of sub-personal *representation*, which I then generalize to organism-level representations, like the outputs of the clocks.

As is the case with any causally-driven information state, the information contained in the outputs of the clocks can be described teleologically.[28] Although my account of the outputs of the clocks does not require a fully articulated theory of teleological content, a few remarks regarding how information from the clocks is used (by animals or humans) in order to obtain information *about relational environmental properties*, suffices to demonstrate that clock representations can misrepresent. I proceed to justify this claim.

Bermúdez explains how sub-personal information can misrepresent in spite of the apparent invariable coincidence that exists between representing and represented structures. What I want to emphasize is the causally driven nature of this information, which in the case of the clocks guarantees isomorphism. All the experimental evidence I used involves representational capacities of animals at the *organism level*, so it is important to emphasize that what I am focusing on are the causal processes involved in these representations, and not their sub-personal nature. Bermúdez's explanation relies on a constraint postulated by Christopher Peacocke (1994), called the 'Overarching Constraint,' which causally driven representations must satisfy in order to have content. Peacocke says that:

> Correct ascriptions of content to subpersonal states are answerable to facts about the relational (environmental) properties of the events they explain, and to counterfactuals about the relational properties of the events they would explain in various counterfactual circumstances (Peacocke, 1994, 312).

In the case of the circadian clock, its phases are answerable to facts about the properties of the period of the rotation of the earth on its own axis and to properties concerning the trajectory of the sun. Moreover, because of the mechanisms for phase-locking, they are also answerable to daily environmental periodic events that are crucial for the animal's survival, as demonstrated by the experiments on time of occurrence discussed in chapter 2. The phases of the circadian clock are also answerable to

[28] See R. Millikan (1984) and C. McGinn (1989) for teleological theories of *content* for sub-personal information.

counterfactuals, such as those concerning the theoretical hypotheses about entrainment and phase drift, also discussed in chapter 2.

Similarly, the stopwatch's outputs are answerable to facts about properties of the environmental intervals they emulate. The metric structure of these outputs allows the animal to predict events, as demonstrated by anticipatory feeding behavior. The constant rate of the stopwatch can be biochemically accelerated or decelerated, which explains various counterfactual circumstances that are experimentally testable, in which behavior anticipating environmental events becomes predictably less accurate.

However, the most important aspect of the relation between the outputs of the clocks and the environmental properties that they represent is that the clocks were *evolutionarily designed* to emulate and represent features of the period of the earth's rotation on its own axis and the duration of intervals of aperiodic events that play a crucial role in an animal's survival strategies. In contrast, tree rings *were not evolutionarily designed* to keep track of the period of the rotation of the earth around the sun. Tree rings *accidentally* happen to register such periods by growing in environments were seasonal changes alter the biochemical composition of the tree. In other words, timing based on the clocks is a capacity *attributable* to an organism as a whole, in virtue of its mental powers to represent time. Clearly, representing time in terms of years is not a capacity attributable to trees (the users of this information have this capacity).

Counterfactuals are very relevant with respect to this issue. Consider the following scenario: suppose that you isolate an information-processing mechanism from the environment. As demonstrated consistently in laboratories, if you isolate an animal from environmental cues, its circadian clock would run free, reproducing the period of the rotation of the earth on its own axis. This is because the evolutionary purpose of the circadian clock is precisely to reproduce such a period. However, tree rings cannot be produced in isolation, i.e., they fully depend on the *combination* of atmospheric conditions that lead to tree-ring formation and the growth of the tree. Thus, tree rings are as much a feature of the environment as they are a feature of the tree. This is why tree rings are so helpful to study ancient climates. The counterfactual situation of isolating a tree to test if its annual clock would run free is simply nonsensical.

Considerations of evolutionary design are very pertinent to teleological explanations of content because they provide the basis for the possibility of misrepresentation in causally driven information processing. Once the functions of the circadian clock and the stopwatch are experimentally identified (as specifying features of the environment that the clocks were evolutionarily designed to emulate), a notion of *proper function* can explain how the clocks may misrepresent. As Bermúdez says,

> Given the particular features that a processing mechanism has been "designed" or "selected" to detect, it is functioning correctly when it responds appropriately to the presence of those features, and incorrectly when it responds in their absence. Correctness conditions are fixed with reference to evolutionary design and past performance (Bermúdez, 1995, 367).

The challenge of demonstrating that the analog representations of the clocks can misrepresent can now be met. Unlike tree rings, the circadian clock and the stopwatch were evolutionarily designed to emulate the period of the earth's rotation on its own axis and aperiodic intervals of important events. The clocks work *correctly* and represent appropriately temporal features when the phases of the circadian clock are locked to the appropriate environmental cues and when the rate of the stopwatch is constant. They misrepresent when the circadian clock free-runs or phase-drifts and when the stopwatch's rate varies. In these cases, the clocks do not invariably respond to the particular features of the environment that they were designed to encode and represent.

Moreover, a seldom noticed point about representations of space and time, emphatically expressed by Reichenbach's notion of 'coordinative definition,' is that these representations inform not by reference to other representations or definitions, but by picking out a specific environmental process. Nor are they conceptualizations or forms of recognitional capacities for specific stimuli. J. J. Gibson (1975) said that, because of the lack of specific temporal stimuli, time cannot be perceived. I suggest that there is a less skeptical conclusion, similar in spirit to Gibson's claim, namely that time cannot be perceived *in the way in which other senses perceive and recognize specific stimuli.*

The uniformity of time and units of time constraints require picking out a reliable process, not recognizing one. This is an extremely important point to bear in mind when mental representations of time are discussed. Finally, as mentioned, counterfactual reasoning (as demonstrated in the time-left paradigm and time of occurrence experiments) is built into the representation space of possibilities produced by the outputs of the clocks.

3.3.5. *Fifth Criterion: Cognitive Integration*

As mentioned, the circadian clock and the stopwatch were evolutionarily designed to keep track of time. In this chapter, I have argued that both mechanisms produce outputs with metric structure that are best described as analog representations. The circadian clock and the stopwatch constitute two different cognitive systems for temporal representation, each with its own type of isomorphism. However, the ubiquitous presence of the circadian clock throughout the spectrum of living organisms seems to call into question whether the circadian clock is really a representational mechanism.

An objection against the representational capacities of the circadian clock could be formulated as follows. What is the status of the outputs of the circadian clocks of plants and bacteria? They seem to have a *proper function*, which comports with teleological accounts of content and misrepresentation. The circadian clocks of plants and bacteria work correctly when they respond to the presence of sunlight and other environmentally relevant features, and they work incorrectly in their absence, e.g., when they free-run. So, are plants and bacteria *representing* metric temporal features periodically?

Setting aside important conclusions from the previous subsection, this objection seems to show that analog misrepresentation is a necessary, but insufficient condition for *analog representation*. Besides the other three criteria (plus analog misrepresentation) a fifth criterion must be satisfied: the outputs of the circadian clock and the stopwatch should admit of cognitive integration. In defining what 'cognitive integration' means in the context of sensory-motor mechanisms, such as the circadian clock and the stopwatch, one must bear in mind that cognitive integration does not necessarily require syntactic structure. As long as there is some type of representational structure, such as *metric structure*, cognitive integration can occur. Bermúdez explains,

> Cognitive integration requires structure, and there seem to be two principal criteria for the presence of structural representational states. First, they must be built up out of components which can be recombined to generate new representational states. Second, the process governing transitions between representational states must be sensitive to their composite structure (Bermúdez, 1995, 365).

As mentioned, symbolic discreteness is not applicable to the outputs of the clocks. However, *both* criteria mentioned by Bermúdez are satisfied by the outputs of the clocks in an analog-metric format. For instance, the capacity to combine outputs of the circadian clock with spatial representations in order to compute the solar ephemeris function satisfies Bermúdez's requirement that structural representational states "be built up out of components which can be recombined to generate new representational states," as does the capacity to combine the outputs of the stopwatch with numeric representations to compute rate.

Bermúdez's other requirement, namely that "the process governing transitions between representational states must be sensitive to their composite structure," is satisfied by the capacity to store the metric outputs of the clocks in memory to register time of occurrence. These memories preserve the metric structure of the outputs of the clocks. Moreover, processes of re-phasing and comparing the duration of intervals, which could be considered as transitions between representational states of the clocks, preserve the isomorphism between temporal representations and the represented periods and intervals. Thus, the metric structure of the outputs of the clocks constitutes a *common code* for computing not only temporal information, but also spatial and numeric information.[29]

In defining what 'cognitive integration' means in the context of the circadian clock and the stopwatch, one must also bear in mind that cognitive integration is an empirical issue, which needs to be confirmed by experimental evidence. In the case of the circadian clock and the stopwatch there is plenty of evidence in support of cognitive integration. Since I covered this material in chapter 2, I will just stress how important the representations of the circadian clock and the stopwatch are for the *sensory-motor system as a whole*. Meck (2003) writes,

> Humans and other animals engage in a startlingly diverse array of behaviors that depends critically on the time of day or the ability to time short intervals. Timing intervals on the scale of many hours to around a day are mediated by the circadian timing mechanism, while in the range of seconds to minutes a different system, known as interval timing, is used. The term *interval timing* is used to describe the temporal discrimination processes involved in the estimation and reproduction of relatively short durations in the seconds-to-minutes range that form the fabric of our everyday existence and unite our mental representations of action sequences and rhythmical structures (Meck, 2003, xvii).

The circadian clock and the stopwatch, as Meck says, are not only critically involved in a startlingly diverse array of behaviors, but they are also critical to *integrate a vast array of information*. Thus, the outputs of the clocks in animals and humans comply with the cognitive integration requirement. Indeed, a significant amount of sensory-motor information integration *depends* on the clocks. With respect to the issue of analog to digital conversion, as long as the *relevant* metric structure is preserved,

[29] For evidence on this common code see S. Dehaene (1997) and C. R. Gallistel (1990).

even though temporal information is lost, metric cognitive integration can include symbolic computations from other areas of the brain by interfacing with metric-analog ones, which is presumably how the ephemeris function is computed.

More importantly, the evidence on spatiotemporal cognitive integration shows that the outputs of the circadian clock and the stopwatch are representations with *content,* because they can be used and interpreted by the agent in such a way that information from the environment contained in these outputs *successfully guides the agent.* In the computation of the ephemeris function, it is the metric mapping of the circadian clock with the period of the rotation of the earth on its own axis that successfully guides the agent toward the fulfillment of multiple goals. This mapping gives *content* to the spatial representation of the position of the sun, because it successfully captures features of the environment (i.e., it satisfies non-trivial accuracy conditions).

There is a fairly direct way of interpreting information from the outputs of the clocks via a function that maps phases of periods or segments of represented intervals with other metric representations, such as spatial coordinates and representations of number.[30] There is absolutely no evidence that plants *use* or *interpret* information from their circadian clock in this way (maybe because they lack a complex nervous system). Because animals and humans use and interpret information from their clocks, the outputs of their clocks constitute representations with content.

To conclude, I take the five criteria I discussed in this section to be necessary conditions for analog representation *only in the specific case of the outputs of the clocks.* Certainly, other weaker and more flexible versions of analog representation are possible. For instance, Blachowicz (1997) says that only relational identity (by which he means some sort of isomorphism) and qualitative or quantitative resemblance are necessary conditions for analog representation. He also says that continuity and density are not necessary conditions for analog representation. This more flexible account might suit other analog representations well, such as those on which Blachowicz focuses, i.e., maps and pictures. But this weaker definition, as I have argued, does not suit the outputs of the clocks well.

---

[30] See F. Egan (1995) for an explanation of the importance of interpretation-functions to determine the content of computations. See Millikan (1993) for how representations that lead to successful behavior can be given a 'consumer semantics' account of their content.

The five criteria I discussed in this section are critical to define the analog representations produced by the circadian clock and the stopwatch. However, one important problem remains to be solved: how does an agent know that a phase of the circadian cycle or a segment of an interval is happening *now*? The representations of the circadian clock and the stopwatch are insufficient to solve this problem. As I will explain in the next chapter, a different type of representation, computed by a different mechanism must be in place. These other sensory-motor representations of time do not have as their main function the preservation of the metric structure of periods and intervals. Rather, their main function is to *locate* the agent in a temporal phase or segment of a represented interval. In other words, these representations constitute the *present moment*.

But before proceeding, I shall briefly mention some of the implications of this chapter for epistemology. The naturalistic approach to epistemology—and philosophy in general aims at answering philosophical questions with views that are at least compatible with the scientific evidence. Chapters 2 and 3 provide a naturalistic account of the basic perceptual-like immediate beliefs (or epistemic entitlements) about time in animals and humans. Moreover, since these chapters demonstrate that these immediately produced beliefs are reliable (because the processes that instantiate the circadian clock and the stopwatch are reliable), they also provide an explanation of why such beliefs are justified (See Goldman, 1992).

John Greco (2010) argues that a virtue epistemology approach to knowledge, understood as success from ability, answers some of the most important questions concerning safety, sensitivity, and the value of knowledge. The abilities that depend on the clocks not only lead organisms to success, but also help them coordinate collective synchronous action. Crucially, chapters 2 and 3 show that an account of basic knowledge about time need not postulate complex self-reflective capacities that involve conceptual and propositional contents involving agency, consciousness, and causation. Reliable abilities based on the metrically-structured representation spaces of possibilities for action and motor control, produced by the clocks, are all that is needed.

CHAPTER FOUR

A TWO-PHASE MODEL OF THE PRESENT (COORDINATION AND EXPERIENCE)

Philosophers have claimed that the present has a special status as a moment in time and have debated about its duration (i.e., does it have duration at all, and if so, how long is it?). A central question is: how should the special status of the present be understood? Some philosophers argue that it should be understood exclusively in psychological terms (the present has a privileged cognitive status, but it lacks a metaphysically privileged status).[1] Others say that the special status of the present is metaphysical and that there are objective facts about the present that explain its unique status.[2]

I will not try to settle this issue here. Rather, I will assume that, at the very least, the present *appears* or *seems* to have a special status. We experience the present in a unique and fundamental way, regardless of whether or not this uniqueness is justified from the perspective of metaphysics. What gives rise to our experience of the present? What in our *cognitive constitution* explains the central role of the present in our subjective experiences? What is the relation between the specification of *simultaneities* for action coordination and the present (i.e., coordination for action and our awareness of the present)? Why does the phenomenal present seem to have duration?

These are questions that can best be answered with the help of cognitive science, which is how I will try to answer them. In this chapter, I postulate a novel *two-phase* model of the present, which offers solutions to philosophical problems regarding the present and is based on experimental evidence from neuroscience and cognitive psychology. A crucial advantage of this model is that it accounts for how the present relates to the outputs of the clocks. According to this model, the present moment needs to be associated with a *sensory-motor* mechanism that interacts with the clocks.

---

[1] See for instance J. J. C. Smart (1963, 1980, 2006), R. Le Poidevin (2007) and C. Callender (2008a).

[2] See for instance R. Chisholm (1990), A. N. Prior (2003) and D. Zimmerman (2005).

Action coordination, the manipulation of information coming from the clocks, and basically any form of interaction with the environment, requires that a particular phase of the circadian clock or a specific moment of an interval must be marked by the mental equivalent of the indexical 'now.' Performing this *indexing* is the task of a sensory-motor mechanism that works independently of the clocks. Certainly, this sensory-motor mechanism is one of the most primitive and fundamental components of the sensory-motor system, and I will describe its main characteristics in this chapter.

The evidence I review addresses traditional philosophical controversies concerning the present, such as: does it have duration? If so, what kind of structure does it have? What is the relationship between the indexical 'now' and the phenomenal present (i.e., the experienced present)? What is the best way to account for the alleged unity and continuity of the phenomenal present? These questions are answered with innovative proposals, the most important of which is the distinction between two types of present, with two different temporal constraints and cognitive purposes (hence the name of the model). The main focus of this chapter is the role of the present in the *representation* of time.

But, one may ask, given that the previous chapters discuss representations of time at length, why devote a whole chapter to the present? Why do so if, with respect to the reliability of the clocks, and the corresponding accuracy for time measurements and mental representations of time, the previous chapters seem to suffice? The answer is twofold, and it requires a brief discussion of the constraints on accurate time measurements and traditional debates about the present in metaphysics and the philosophy of mind. With respect to the first issue, concerning general constraints on time measurements, Reichenbach (1958) said, in his philosophical analysis of space and time:

> After we had specified the unit of time, which is the first metrical coordinative definition of time, we were led to the problem of uniformity, which is the second metrical coordinative definition of time and deals with the congruence of successive time intervals. There is however a second type of comparison that concerns parallel time intervals occurring at different points in space rather than consecutive time intervals occurring at the same point in space. The comparison of such time intervals leads to the problem of simultaneity and hence to the third metrical coordinative definition of time (Reichenbach, 1958, 123).

As mentioned, the reliability of the clocks guarantees that identical phases and segments of intervals are equal in duration, which can be used as units of time because the cycle of the circadian clock and the rate of the stopwatch are *uniform*. But the clocks cannot be used to determine whether two things that happened at different locations occurred at the same time. The synchrony or asynchrony of two events must be specified independently. In physics, this leads to a coordinative definition of simultaneity, which involves clocks at two locations, distances, and signals (for instance concerning light).

But, obviously, for organisms that use the circadian clock and the stopwatch to represent time, their sensory-motor system must somehow determine simultaneous events without having clocks at *different* locations. The sensory-motor system performs this task by using a remarkable set of integration mechanisms called 'simultaneity windows.' Each of these windows is very well tuned to the type of speeds involved in different sensorial stimuli (e.g., light versus sound). The scientific research on simultaneity windows has revealed that the human brain is capable of specifying simultaneity with great accuracy, albeit in an approximate way. However, very little attention has been devoted in the literature to the important theoretical issue of how to understand the interaction between different simultaneity windows and the clocks. A major goal of this chapter is to provide an account of such interaction.

With respect to debates about the present in metaphysics and the philosophy of mind, it should suffice to say that what gets discussed in some of these debates is entirely different from the coordinative definition of simultaneity required for accurate time measurements, because these debates mainly concern our subjective awareness of the present moment. Likewise, the issue of how our awareness of the present relates to simultaneity windows has not received a careful assessment based on empirical findings. I offer an account of this relation in the second part of this chapter.

The structure of this chapter is as follows. In section 4.1., I review the psychological evidence on simultaneity windows and postulate the first component of the two-phase model of the present: the sensorial present. In section 4.2., I describe the main cognitive function of the sensorial present, which is to anchor the outputs of the clocks. In section 4.3., I discuss the main philosophical issues concerning the conscious awareness of the present, review the experimental evidence on the conscious present, and conclude by postulating the second component of the two-phase model: the phenomenal present. Finally, in section 4.4., I outline the main characteristics of the two-phase model in more detail.

4.1. Simultaneity Windows and the Units of Time: The Third Constraint on Coordination in Time

One of the most important and robust psychological findings concerning the present is that there is a measurable amount of time during which stimuli are *judged (or registered) as simultaneously present* by the sensory-motor system, even though there is an interval separating the stimuli.[3] These unconscious judgments for motor control and action coordination are used in unison with the accurate time measurements of the cognitive clocks. As I explain in the following subsection, the brief amount of time that separates the stimuli, which is called a 'simultaneity window,' *varies* across modalities and depends on several factors, including age and specific properties of the stimuli. But before describing these and other interesting aspects of simultaneity windows, I shall explain their relationship to the clocks.

As mentioned, the circadian clock and the stopwatch are extremely efficient registers of temporal information. However, the simultaneity of sense stimuli needs to be established *previously*, in order for the clocks to determine the amount of time that separates non-simultaneous stimuli. Thus, simultaneity is crucial to specify units of time that are environmentally indispensible for decision-making and navigation (e.g., is the sound I am hearing simultaneous with the thing I am seeing?). Once the simultaneity of two events is determined, the nervous system can specify what phase or interval corresponds to it. The clocks by themselves cannot solve this problem (they are always at a single location at a time: the agent's location) and using multiple clocks at different locations to compute simultaneities, as they are computed in physics, is not an option for the nervous systems of creatures that represent time.

Determining the simultaneity or non-simultaneity of radically different sensorial stimuli also presents computational challenges about *speed*

---

[3] I will use the term 'judged simultaneity' instead of the more common term 'perceived simultaneity' because it avoids problems concerning the content of *specific* perceived stimuli (there are no specific stimuli for simultaneity). Also, identifying a perceptual organ for simultaneity is problematic. In contrast, 'judged simultaneity' is neutral with respect to these issues and correctly conveys the notion that these are cognitive processes in which stimuli are *taken to be simultaneous* by the sensory-motor system, depending on different circumstances. Obviously, these judgments happen at the motor control level and do not require conscious inferences of any kind. They are the result of functional procedures that produce an outcome: e.g., perceptual stimuli *a*, *b*, and *c* are simultaneous. See H. v. Helmholtz (1873) for the origins of this notion.

that cannot be solved by the clocks. For instance, light travels about a *million times* faster than sound. Determining the simultaneity or non-simultaneity of visual and auditory information depends on specific characteristics of the visual and auditory systems, and on the distance between the source of information and the receiver. Thus, the most efficient way to determine the simultaneity of sensorial stimuli is that *each sense modality* should have its own *threshold* of simultaneity. The evidence confirms not only that this is the case, but also that there are processes that compensate for differences in simultaneity judgments from different senses concerning the same stimuli sources. So the first cognitive step for calculating the duration of intervals or periods involving multiple stimuli from different locations is to determine simultaneities, which can then be *correlated* with segments of intervals or phases of periods.

This section is structured as follows. I first review the psychological evidence on sense-specific simultaneity windows. Second, I discuss evidence suggesting the existence of a cross-modal simultaneity window. Third, I explain the relevance of the cross-modal simultaneity window by highlighting its relationship with the clocks. Finally, I conclude by defining the most important properties of the cross-modal window and characterize it as the *sensorial present*, which is the first component of the model that I defend in this chapter.

4.1.1. *Sense-Specific Simultaneity Windows*

As mentioned previously, the optimal solution to determine the simultaneity or non-simultaneity of stimuli is that each sense modality should have its own window of simultaneity because of the different characteristics of the media that the senses register (e.g., light and sound). In this subsection I present experiments which confirm that indeed each sensory modality has its own window of simultaneity. For instance, the auditory window of simultaneity can be identified by the so-called 'click-fusion' experiment, in which two tones or clicks that are separated by an interval of 2 ms are experienced as a single tone or click. Ernst Pöppel explains,

> In the case of these chronological differentials, one hears always only one tone, even when an objectively measurable differential, for instance of two thousandths of a second, exists between the two stimuli. The objective chronological difference is in other words insufficient to produce the experience of two separately heard tones. What is separated by two thousandths

> of a second, what is objectively non-simultaneous, appears subjectively as *one* event, that is to say: In the case of these two acoustical stimuli, we find ourselves inside a single "window of simultaneity" (Pöppel, 1988, 12).[4]

Pöppel goes on to explain that the threshold for registering stimuli as simultaneous varies across individuals. The auditory simultaneity window can vary from 2 to 5 ms, and there are variations that depend on properties of the stimuli, such as loudness. There is also an age/value correlation: the older the person, the higher the value of the simultaneity window (Pöppel, 1988, 12). This means that for some people, stimuli can be separated by 5 ms intervals and still be registered as only one stimulus. Once one crosses this threshold (2–5 ms), the stimuli are correctly registered as two distinct sounds.

Thus, the same stimuli, say two sounds with the same pitch, separated by an interval of 4 ms, may be judged as simultaneous by one person and as non-simultaneous by another person. The reason why these differences do not cause inter-subjective chaos is that at such small scales, variations in judged simultaneity are negligible. Nonetheless, such differences are undeniable and surprising because two distinct sounds are judged to be, within the simultaneity window, a single sound. Equally surprising is the fact that different sense modalities have substantially *longer* simultaneity windows. To quote Pöppel once more,

> If one stimulates the skin with stimuli of short duration, the window of simultaneity is enlarged to about ten thousandths of a second [...] If a similar experiment is carried in the visual modality, sight, we obtain yet another result. In the neighborhood of twenty to thirty thousands of a second have to elapse before two visual impressions appear as nonsimultaneous. Below this temporal boundary everything is simultaneous. Although we like to characterize ourselves as visual animals, our visual system, compared to hearing or touch, is very slow (Pöppel, 1988, 16).

Pöppel suggests that the lack of experiments on simultaneity windows in the other two senses, taste and smell, may be explained by the technical difficulties for measuring such windows. But with respect to vision, audition, and touch, experiments on simultaneity windows have confirmed the bizarre fact that the same temporal interval between stimuli can lead to incompatible judgments of the stimuli as simultaneous or non-simultaneous, depending on the sense modality.

---

[4] Although Pöppel uses the word 'experience' it will become clear that the processes involved in judgments of simultaneity are unconscious and lead to motor control without conscious monitoring.

The substantial differences in the duration of simultaneity windows (vision has a simultaneity window *ten times larger* than audition) generate a very significant problem. Many auditory and visual stimuli come from the same source. The brain needs to identify the spatiotemporal location of the source and parse the stimuli into simultaneous and non-simultaneous in order to be able to coordinate successful action. But with the discrepancies between simultaneity windows just mentioned, one may ask, how are the differently-parsed stimuli integrated into a *single sensorial window*, where stimuli from *all the senses* are judged as simultaneously present? Answering this question is the main purpose of the next subsection.

4.1.2. *The Multi-Sensory Integration Window*

The evidence on simultaneity windows reveals that there are substantial differences in their duration, which creates the problem of explaining how it is possible to perform sensory-motor tasks that require cross-modal integration of stimuli into a single multi-sensory percept, in virtue of which stimuli from different senses are judged to be simultaneous. As J. V. Stone, et al. (2001) explain in the following quotation, multi-sensory simultaneity is one of the most fundamental requirements for motor coordination, and there are two possible explanations for how multi-sensory simultaneity may be computed by the brain:

> When executing time-critical tasks, such as playing table tennis, knowing precisely when the ball made contact with the table is important for fast and accurate motor coordination. However, even if the perception of audio-visual simultaneity is not veridical, it should at least be stable for a given observer. Such stability may permit the motor system to be temporally calibrated with respect to the perceived timing of auditory and visual events. These considerations suggest that the perceived timing of visual and auditory events should be highly accurate, or, at least, highly stable for a given observer (Stone, et al., 2001, 31).

Stone and his collaborators tested these two hypotheses, focusing on the audio-visual integration of stimuli. The first hypothesis—that the judged simultaneity of stimuli depends on a highly accurate mechanism—predicts that differences in judged simultaneity across individuals are very unlikely to occur. If this is false, then the second hypothesis predicts that there should be a high degree of stability in the duration of the integration window for audio-visual judged simultaneity in each individual, but that there will be variance across individuals.

Stone, et al. corroborated the second hypothesis. Through a series of experiments that dissociate reaction time tasks (RT) from the so-called 'point of subjective simultaneity' (PSS), they obtained the following results. First, the PSS values, which reflect the most likely threshold at which an observer will judge audio-visual stimuli as simultaneous, are *observer specific*.[5] Actually, each observer's PSS value is significantly different from other observers and from the estimated population mean (see Stone, et al., 2001). This finding shows that there is variation across individuals with respect to critical aspects of the *multi-sensory simultaneity window*, such as the PSS value. But the PSS value is stable over time for each observer.

The PSS value depends on *spatial* information (which is not surprising given that, as mentioned, simultaneity judgments depend on the time of events at different locations). For instance, suppose that the PSS is 50 ms, which means that if the onset of the light stimulus comes 50 ms before the onset of the sound stimulus, then the sound and light stimuli onsets are judged as simultaneous.[6] This particular PSS of 50 ms depends on the *location* of the observer relative to the source of the stimuli. More specifically, the PSS depends on a *horizon* of simultaneity where the distance of the source of stimuli from the observer, given the radically different speeds of light and sound, is critical for the specification of the numeric value of the PSS. Anatomical features of the sense organs, in this case audition and vision, also play an important role. It turns out that, once one takes into account all these factors, the horizon of simultaneity for light and sound is about 10 meters from the subject.[7]

But what happens when the subject is not exactly at 10 m from the source of stimuli? How does the brain compensate for differences in synchrony, due to the subject's location or other causes of asynchrony, in order to determine that the stimuli occurred simultaneously? As Craig Callender (2008a) says, although there will be a point at which the horizon of simultaneity is far enough from the subject to hinder judged simultaneity, the sensory-motor system's mechanism for determining the

[5] More precisely, the point of subjective simultaneity is the *stimulus onset asynchrony* at which an observer is *most likely to judge the onset* of a light and a sound as simultaneous.

[6] The Stone, et al. experiment is based on stimulus onset asynchronies that range from −250 ms (sound before light) to +250 ms (light before sound). This 250 ms window is very relevant, as the research that I am about to discuss reveals.

[7] See C. Callender (2008a) and references therein.

simultaneity of cross-modal stimuli is very tolerant to asynchronous information. Callender explains,

> At some point the brain does not weld the two aspects of the event into a simultaneous whole. The phenomenon of thunder and lightening is perhaps the most conspicuous such case. We hear the sounds later than seeing the light. And if the event is up close, we can react quicker to an auditory source than to a visual one; so up close there are cases where the brain—at least for quick reactions—is not waiting for the visual processing to catch up. Still, the brain is surprisingly tolerant of asynchronous information. There are no noticeable discrepancies between the image of the lips moving and the sound "Now!" at any typical communication distance (Callender, 2008a, 14).

Callender is referring to an experiment by Dixon and Spitz (1980), which shows that the sound of the word 'now' could be produced as much as 250 ms *after* a person moved her lips (saying 'now'), without the subjects noticing the discrepancy. It is remarkable that the multi-sensory window of integration, in this case limited to audio-visual stimuli, is as large as 250 ms. Although PSS values, which are shorter than the window of 250 ms (because they are the threshold at which a subject will *most likely* judge auditory and visual stimuli as simultaneous) vary from subject to subject, the window for multi-sensory information seems to be around 250 ms across subjects. Indeed, the temporal window used by Stone, et al. to identify the PSS values in their study is 250 ms.

A wide multi-sensory integration window of 250 ms (a window that is about a *hundred times larger* than the 2 ms window of simultaneity for auditory stimuli) explains the surprising tolerance of the brain to asynchronous information. Furthermore, other experiments show that there is an active process of *adjustment* and *recalibration* that the brain must perform for integration to happen. Some of the most relevant experiments concerning this issue include those on temporal ventriloquism (e.g., Welch, et al., 1986; Spence and Squire, 2003), motor-sensory recalibration regarding the effects of the subject's intentions (e.g., Eagleman and Holcombe, 2002; Stetson, et al., 2006), cross-sensory recalibration (e.g., Fujisaki, et al., 2004), and distance-based recalibration (e.g., King, 2005).

It is plausible to hypothesize that there is a wide simultaneity window of about 250 ms for multi-sensory integration. Although the experiments I have mentioned in this subsection concern exclusively audio-visual stimuli, it is reasonable to postulate that a similar window would be required for the brain to recalibrate multi-sensory stimuli, in order to compensate for asynchronies across the sensory modalities.[8]

Experiments on the multi-sensory integration window are ongoing. It is a fascinating area of research that is revealing the level of complexity underlying the most basic acts of perception. Two fundamental problems, however, are generated by the multi-sensory integration window. One concerns the relationship between the clocks and the integration window. The other is the nature of the relationship between the integration window and the shorter, sense-specific simultaneity windows. These are the issues that I address in the next subsection.

4.1.3. *The Multi-Sensory Integration Window and the Clocks*

There is another temporal threshold that is highly relevant to understand the relationship between the integration window and the clocks, particularly the stopwatch.[9] I will call this threshold, following Pöppel, the *order* threshold, because it allows subjects to determine which of two non-simultaneous stimuli came first. Below this threshold, subjects can *distinguish* the stimuli, but they are unable to determine *which came first*. In the case of auditory stimuli, which have the smallest simultaneity window, experiments have shown that there is an interval of about 25 to 35 ms. Within this window, subjects can distinguish two sounds, but are unable to determine which sound came first. Pöppel describes the results of these experiments as follows:

> A person is in a position to give accurate information only when the chronological separation between the two tones lies in the neighborhood of 30 to 40 thousandths of a second. Although two distinct tones can be heard, a period ten times as long as the approximately four thousands of a second has to elapse before certainty can exist as to which was the first tone and which the second (Pöppel, 1988, 19).

A very important characteristic of the order threshold is that it is *not* sense specific. This is a perplexing feature of the order threshold because one would think that perceiving the order of sensorial stimuli is something that happens at the lowest level of information processing. Certainly, one would expect that by distinguishing two stimuli as two distinct sounds,

---

[8] I will soon characterize the mechanism responsible for integration and recalibration within the 250 ms window in more detail.

[9] I say 'particularly the stopwatch' because the stopwatch depends on sensorial stimuli for its activation. But since the window I am about to discuss determines the order of sensorial stimuli, it is also relevant to determining the phase information of the circadian clock.

the auditory system would *immediately* register which sound came first. However, experiments have shown that the discrimination of stimuli is a necessary, but not sufficient condition for chronological order perception. Why is the chronological order of cross-modal stimuli determined after a *fixed* threshold for all modalities if the sense specific windows are *substantially different*? Indeed, it seems very counterintuitive to think that a sense-*independent* threshold should determine the order of *sense-specific* stimuli. But this is exactly what researchers have found. Pöppel explains,

> When the experiment is carried out with tactile or optical stimuli, a marked differential from the window of simultaneity is found in these sensory systems as well. To be able to say that something came first or second requires interestingly enough in each case the same time interval for the three sensory systems [...] viz., approximately 30 to 40 thousands of a second, whereas the span of simultaneity is in each case totally different, as we had discovered (Pöppel, 1988, 19).

Where does this order threshold come from? Notice that in the case of vision, the simultaneity window of 20 to 30 ms and the order threshold jointly determine an almost *immediate* discrimination of stimuli *and* their chronological order because there can be at most 20 ms between the discrimination of stimuli as non-simultaneous and the specification of their chronological order. However, the fact that there is a fixed 30 to 40 ms threshold for cross-modal stimuli suggests that there is a *common mechanism* that *compensates* for the differences in durations of the sense-specific simultaneity windows and *integrates* stimuli in terms of chronological order across the modalities. Now, the question is what are the characteristics of this mechanism? Specifically, is it a clock or a non-temporal mechanism?

There are two hypotheses that have been discussed in the literature, both of which postulate a specific type of relationship between the clocks and the cross-modal integration window. The first hypothesis postulates that a *clock* is the mechanism responsible for the temporal integration of sensorial stimuli into an ordered sequence. The second hypothesis postulates that there is a non-temporal sensory mechanism that integrates cross-sensory information into ordered sequences that can *then* be used for clock-dependent action coordination. I proceed to explain these hypotheses and argue in favor of the second hypothesis.[10]

---

10 Notice that the issue is not how the *simultaneity* of stimuli is determined. Both of these hypotheses *assume* that there are windows for judged simultaneity. The question is, what is the mechanism responsible for determining the *chronological order* of stimuli?

As explained in chapter 2, interval timing is the capacity to estimate durations in the seconds to minutes range. If the stopwatch can discriminate durations at the seconds range, could it also have the capacity to discriminate durations at the milliseconds range? According to the first hypothesis, the order threshold could be explained as the minimum amount of time required for information to be encoded in the memory component of the stopwatch. This explanation has the advantage of being parsimonious, because it avoids postulating a non-temporal cross-modal mechanism for chronological order that outputs information into the clocks. The order threshold is fixed and cross-modal because it is a property of the stopwatch's memory component.

Although this is a plausible explanation of the fixed value of the order threshold, evidence favors the second hypothesis. Karmarkar and Buonomano (2007) developed a neural network model for timing at the milliseconds to seconds range that encodes time-dependent information by identifying spatial patterns of neural activation independently of any clock-related information. Ivry and Schlerf (2008) describe the experimental evidence in support of this model as follows:

> Secondary tasks or pharmacological manipulations affect judgments of 1 s intervals while having little or no effect on intervals of around 100 ms. Secondary tasks that affect judgments of 1 s have little effect on intervals of 100 ms. Temporal acuity normalized to mean duration is relatively constant for intervals between 200 ms to 2 s but becomes considerably poorer for intervals shorter than this range (Ivry and Schlerf, 2008, 275).

These results suggest a significant discontinuity between the milliseconds and the seconds range, evident at around 100 ms, which is well above the order threshold. Additionally, the architecture of the sensory-motor system seems to require a non-temporal mechanism, as the second hypothesis postulates, because sensorial information is also relevant for determining the phases of the circadian clock. Given the differences between periodic and interval timing, it is more efficient for the sensory-motor system to have a non-temporal mechanism that outputs sensorial information to the clocks, rather than to integrate such information through the stopwatch.

Moreover, the general properties of interval clocks (particularly the scale/resolution tradeoff), which apply to the stopwatch, make it an inadequate mechanism for temporal registration at such a short scale. As Ivry and Schlerf say: "It is unlikely that a single mechanism could operate at these different time scales. A pacemaker used to judge an interval of 40 s is unlikely to have the resolution to judge a 100 ms interval" (2008, 275).

The second hypothesis should, therefore, be accepted as the correct explanation of the threshold for chronological order, which occurs *within* the range of the multi-sensory integration window, i.e., about 250 ms.

To conclude, this cross-modal mechanism is not designed to register or represent time beyond the milliseconds range and it integrates sensorial information into *ordered* sequences of stimuli, which are *previously integrated* by sense-specific simultaneity windows. The clocks rely on the sensorial information integrated by this mechanism. Although this integration mechanism has temporal *constraints*, as evidenced by the order threshold, it is not a clock. Thus, the order threshold seems to be a property of the integration window, and not of the stopwatch. In the next subsection, I address the issue of whether the multi-sensory integration window satisfies the characteristics of what philosophers have called the *specious present*.

4.1.4. *The Sensorial Present*

In this subsection, I will argue that the multi-sensory integration window has some of the characteristics that philosophers have traditionally attributed to the specious present. I shall assume that this simultaneity window is the relevant one for action-coordination at the *organism level*, and therefore, that it should be characterized as the fundamental *representation of simultaneity for motor control*. Should this simultaneity window be characterized as the specious present?

Since its introduction to the philosophical literature by William James, the characterization of the specious present has focused on its *duration* (e.g., the duration of short-term memory or the duration of a unit or event, such as a sentence or a musical phrase). These characterizations also emphasize that the specious present has a structure that links past stimuli with future ones, which is allegedly the basis for the perception of motion and change. In contrast, the defining characteristic of the cross-modal simultaneity window is that stimuli are judged as simultaneous by the sensorial system, *regardless* of specific durations.

One could stipulate, however, that the specious present is the cross-modal window because it is not a durationless instant. It is very important to emphasize that 250 ms is the *maximum* duration of the cross-modal window. This window, which I shall call the 'sensorial present,' happens to be the longest window for judged simultaneity, but this does not mean that non-simultaneity cannot be registered within much shorter durations, such as the 2 to 5 ms window of the auditory modality, at the *sub-personal* level. The cross-modal window provides a representation of simultaneity at the organism level (a representation of cross-modal stimuli that are *not* simultaneous *as* simultaneous). Since this window is not a durationless instant, but always a brief interval, it is a specious kind of present in the sense James intended.

There is an important issue that requires clarification before proceeding. These representations of simultaneity are fundamental for motor control and sensorial calibration at the organism level, but motor control occurs *unconsciously* (see D. A. Rosenbaum, 2002). Therefore, it would be a mistake to characterize these representations of simultaneity in terms of contents of which one has phenomenal consciousness (conscious experiences as of simultaneity, or simultaneity *seemings*). Likewise, such information cannot be consciously monitored or introspected. This has not been emphasized in discussions about the specious present (as it relates to time representation and consciousness). A proper understanding of this issue, which is based on psychological findings on motor control, reveals that central aspects of the philosophical debate on the specious present may be based on a mistake.

To illustrate this point, think of syntax processing and other motor-control aspects of linguistic representation. Syntax processing consists in the systematic manipulation of information in terms of strictly formal characteristics. A string of symbols or sounds are processed according to formal distinctions such as subject, predicate, noun, adverb, etc. In the case of speech production, one is conscious of the meaning of words, but not of the articulatory code that is involved in gesticulation and sound intonation. Similarly, one is consciously aware of the meaning of sentences, but not of how their syntax is processed by the brain. This distinction between conscious experience and unconscious access to information should inform the characterization of any kind of behavior in which there is a motor-control component and an action-selection component with phenomenal perceptual features.

For instance, consider the case of a hypothetical baseball player, Anna, who is about to hit the home run that will help her team win the game. What is going through her mind? A lot, one may say: she may be thinking about the score of the game, the inning, and the need for a home run. She may also reflect on her feelings—a combination of expectation, excitement and anxiety. Some other thoughts may cross her mind, like how being in such a situation reminds her of an episode during her childhood. It is difficult to tell how much information she is consciously experiencing at any specific moment. But there are two aspects of her cognitive situation that can be determined without much controversy.

The first, as mentioned, is that motor control is unconscious. Anna may be thinking about a lot of things, but she has no introspective access to the extremely precise representational system that allows her to hit the home run. As she swings the bat, she has no access to the motor system that calculates the speed and trajectory of the ball to such a high degree of precision that her fast swing coincides with the ball's trajectory. The motion and speed of her arm, the position of the bat, the speed of the ball and her overall posture contribute to this remarkable feat. All this happens as a result of complex representations and calculations about time, space, speed, etc. But Anna is unconscious of the precise speeds at stake, the calculation of speeds and trajectories, the precise position of her body, and how her position adapted during the swing. The bat hits the ball and the ball leaves the field. Anna is certainly aware of this perceptual information. She paces along the field, but again, she has no access to the motor system that allows her body to move smoothly across the field without falling down or bumping into objects or other players. She might be extremely emotional, but this does not hinder her motor control as she runs. Her phenomenal consciousness at that moment involves a set of experiences, including excitement. However, the motor-navigational system that allows her to run around the field is far beyond her conscious reach (fortunately, because otherwise, Anna would be severely slowed down by distracting information concerning multiple speeds and calculations). She just *performs* these motor tasks (in the case of hitting a home run, a particularly remarkable one), without really being aware of *how* she succeeds in performing them.

Unconscious processing is characteristic of any type of motor control for automatic behavior (e.g., violin playing, tango dancing, bicycle riding) and of motor control in general. What is particularly relevant for the topic of this chapter is that *all* motor control requires temporally organized information, packaged unconsciously into cross-modal simultaneity windows specified by the sensorial present.

The second aspect of Anna's cognitive state that can be established with a high degree of certainty concerns the stability of the experiences of which she is conscious. Her excitement is part of a set of experiences that are unified in a very robust sense. How is this unity possible at the conscious level and how does it relate to the non-conscious representation of simultaneity for motor control? I will address questions concerning the temporal aspects of conscious awareness in section 4.3. What I want to emphasize now is the importance of the distinction between unconscious motor control and conscious forms of perceptual representation. This distinction is crucial to avoid misunderstandings concerning judged simultaneity at the sensorial level, and apparent simultaneity and the experience of succession at the phenomenal conscious level.

Many factors contribute to the representation of sensorial judged simultaneity. Some of the most important are: speed of the medium that is registered (e.g., light, sound, etc.), distance from the source, and time between detection of the sensorial signal and brain processing. Using a traditional example, there are several times involved in the perceptual act of seeing a cosmic event that happened a long time ago (e.g., the times of the occurrence of the event and of the sensorial registration). What matters for accurate motor control, however, is not whether *all* these different times are accurately represented. Rather, and in accordance with Reichenbach's notion of coordinative definition, one must *stipulate* what counts as simultaneous for the sensorial system at the organism level. Obviously, in the case of motor control, it is the brain (specifically the motor-control system at the organism level) that stipulates such definition: this stipulation *is* the sensorial present.

A high degree of accuracy and reliability is involved in such stipulation of simultaneity, and the complexity of calibrating action and automatic responses while registering stimuli from different media at different speeds is remarkable. As mentioned, reliability is critical for epistemic success and knowledge, because it depends on facts about the environment that determine rates for error avoidance. This is why a scientific account of the sensorial present should constrain and inform philosophical debates.

Philosophers are currently debating the adequacy of different models for temporal experience and forms of specious present, but none of these models is strictly based on scientific findings. Rather, these models appeal to metaphysical distinctions, such as mereological relations that hold between wholes and their parts (e.g., a whole present experience and its parts at a time). There is also intense debate concerning how best to understand the distinction between a succession of experiences and the experience of succession. The three models offered in the literature (the cinematic, retentional, and extensional models), appeal to conscious experiences without making explicit reference to psychological findings.[11]

---

[11] The cinematic model characterizes conscious experiences as the result of the succession of temporal "slices" of simultaneously integrated information, similarly to the way in which the rapid succession of a set of pictures produces the appearance of motion in a movie. The retentional model adds 'retentions' to the cinematic model, claiming that recent slices (the retentions) are always consciously integrated with the slice that is

Instead, most of the debate is based exclusively on introspection, its limits, and the plausibility of some type of metaphysical relation for experiences.[12] This methodology has two important shortcomings.

First, one may worry about the very subject matter of these debates. For instance, one may suspect that there are no experiences at all, or that their metaphysical existence is quite problematic.[13] For how should experiences be individuated and how could one quantify over them? Presumably, one of the main criteria for their individuation would be their temporal extent, but then substantial problems emerge. Given the extremely short duration of the sensorial present (provided one agrees with the findings on motor control) there would be an explosion of metaphysically complex structures with subparts (whole experiences and their sensorial and phenomenal components). Suppose instead that one allows experiences to extend to whole unified streams of consciousness (whole days in wakeful consciousness and whole dreams, for instance). Then what prevents us from saying that a whole life is a single experience? I revisit this problem in section 4.3.

Second, and more importantly, the methodology used in these philosophical debates about the specious present (e.g., a priori reasoning, introspection, metaphysical criteria for individuation) is *incompatible* with the scientifically constrained methodology that Reichenbach (1958) rightly advocated to address philosophical problems, particularly about space and time. For example, nowhere in these debates can one find a model that carefully distinguishes simultaneity representations for unconscious motor control and clock representations from temporal representations in conscious experiences.

In order to offer a scientifically informed account of the specious present that takes into consideration this distinction, I propose that one should interpret these models for the specious present not as rival characterizations of the same phenomenon, but rather, as *answers to different questions*. One can illustrate this point with Ned Block's (2003) suggestion that physicalism and functionalism about consciousness may not be rival theories, but answers to different questions. According to this suggestion, physicalism would address issues regarding the neural basis of conscious

experienced as present. The extensional model hypothesizes that conscious experiences have a temporal extension that cannot be captured in terms of temporal slices. I elaborate on these models below.

12 For a thorough assessment of these models see B. Dainton (2010).

13 For a defense of the inexistence of experiences in the robust metaphysical sense assumed in debates about temporal experience, see A. Byrne (2009).

experience, while functionalism would attempt to answer questions about the nature of neural representations that are available for thought, decision, reporting, and action.

Similarly, I suggest that extensional models try to account for our conscious experience of succession, as well as the continuity and unity of conscious experiences, while the cinematic and retentional models try to answer questions about how information concerning represented simultaneities for motor control become available for thought, decision, reporting, and action. In other words, extensional models are concerned with what I will call the phenomenal present, while cinematic and retentional models are concerned with the sensorial present and its relation to access consciousness. The remainder of this chapter is devoted to clarifying and explaining this distinction.

An indication that the present proposal may be on the right track is the commitment of retentional and cinematic models to the principle of simultaneous awareness. Representations of simultaneity are crucial for navigation, and simultaneity is a fundamental constraint on time measurements. It seems that the question that retentional and cinematic models try to answer is, how does information about represented simultaneities become available for thought, decision, reporting, and action? Psychological findings have demonstrated that this information is available for decision-making and action *without* conscious awareness. It is, therefore, important to explain how information about represented simultaneities can become available for thought and reporting without assuming conscious awareness.

The sensorial present seems to be the kind of specious present postulated by retentional and cinematic theories. These unconscious representations of simultaneity, like the outputs of the clocks, are susceptible of cognitive integration and interpretation, along the lines suggested by these models. If I am right about this, then the whole debate on temporal consciousness and the specious present needs to be reformulated. What is important at this point is to describe the representational *role* of the sensorial present in navigational tasks and motor control.

A crucial question that needs to be addressed in order to understand the importance of the sensorial present is: what happens to the asynchronous information from sense-specific windows when the sensorial present integrates it cross-modally? The evidence shows that asynchronous sense-specific stimuli are judged as simultaneous within a *single* window of simultaneity for cross-modal stimuli. This means that the sensorial present imposes the relation of simultaneity among asynchronous information, as it were, a *second time*. The "first" sense-specific judged simultaneities unify asynchronous stimuli into a simultaneous whole, but unlike the sensorial present, this process of unification is not one of *cognitive integration* across modalities.[14] Furthermore, as mentioned, the sensorial present is the relevant representation of simultaneity for cognitive integration at the organism level. This explains why even the asynchronies determined by the order threshold within a sense modality are ignored by the mechanism responsible for determining the sensorial present. This can be captured by the following principle:

> **Simultaneous Unification:** Once the asynchronies of modal specific windows are integrated and re-calibrated into the sensorial present, these asynchronies disappear and the stimuli are interpreted by the motor-control system as falling under a single moment in which cross-modal stimuli are related as simultaneous.

Retentional and cinematic models of temporal consciousness assume an analogous principle (the *principle of simultaneous awareness*). According to this principle, conscious mental contents, in order to be unified into a single experience, must be simultaneously presented to a single momentary awareness (see Dainton, 2010). If I am right in my assessment, this type of consciousness is only *access* consciousness, which complies with the functional role of making information globally available for different cognitive tasks.

Stimuli registered within the sensorial present are indisputably related by *simultaneity*, which is a *symmetric*, *transitive*, and *reflexive* relation. Judging something as chronologically ordered requires an *asymmetric* relation, e.g., 'before than' or 'after than.' This is why the order threshold is ignored in the representation of the sensorial present. Clearly, stimuli are either related by simultaneity or by an order relation, they cannot be related by both.[15]

What *kind* of mechanism is responsible for integrating stimuli into simultaneities? I mentioned that the mechanism responsible for determining the sensorial present is a non-temporal mechanism. But this

---

[14] For the importance of cognitive integration as a criterion for representation see chapter 3, specifically 3.3.5.

[15] What about the perception of change and motion? Perceiving change and motion depends on mechanisms that *differ* from those involved in the integration of stimuli in the sensorial present, which only concerns the *simultaneity* of stimuli. However, perceiving change and motion *depends* on the sensorial present, the clocks and spatial representation mechanisms.

negative characterization is vague: many mechanisms are non-temporal. In the next section, I argue that however it is instantiated and however many sub-components it may have, the mechanism responsible for sensorial present representations is a *meta-semantic mechanism*, and also that the sensorial present is the *temporal mental demonstrative*, the mental equivalent of the indexical 'now,' for the purposes of motor-control (sensory-motor) representation (e.g., navigation, trajectory calculation, etc.).

4.2. Meta-Semantic Mechanisms and the Temporal Indexical

The sensory-motor system needs to coordinate action-oriented representations with features of the environment (particular regions of space and time, and particulars that can be *demonstratively referred to* in general). Thus, it needs to *individuate* and select packets of information that are relevant for action, perception and motor control, and it also needs to eliminate noise or unnecessary information. Moreover, the analog representations of the clocks need to be *anchored* to the sensorial present in order to integrate accurate temporal information for cross-modal sensory-motor representation. The sensorial present is determined by a finely-tuned calibration mechanism. The relationship between the calibration mechanism and the clocks is that this mechanism individuates the packets of information that the motor control system uses to anchor the analog representations of the clocks, thereby reliably satisfying the three constraints on accurate time measurements.

More specifically, circadian clock representations need to be anchored to the sensorial present for navigation and other motor control tasks. For instance, computing the solar ephemeris function depends fundamentally on determining which phase of the clock corresponds to the present moment. Phase resetting also depends on such anchoring because it basically consists in changing the phase that corresponds to the present. Likewise, the representations of the stopwatch need to be anchored to the sensorial present in order to specify time elapsed from the moment of activation.

One can easily exemplify this kind of anchoring by comparing the clocks with a speedometer, which was a useful way to illustrate the analog nature of clock representations (the main topic of chapter 3). Speed is continuously represented by the surface on which the pointer of the speedometer moves. The pointer moves continuously and smoothly through the surface, always marking a specific speed. Analogously, the clocks need a *pointer that selects* a particular phase of the circadian cycle or a specific moment of an interval registered by the stopwatch. This pointer is the sensorial present. Like an arrow on a map, the sensorial present informs the sensory-motor system what phase or interval is occurring now. But what kind of *representation* is the sensorial present, such that it can anchor the representations of the clocks? It is the sensorial representation of cross-modal simultaneity, but this characterization does not suffice to explain what type of *mental representation* the sensorial present is.

The sensorial present is a representation of simultaneity in the sense that it is a sensorial *construct*. But if the sensorial present is a representation, how can it *misrepresent*: it is not clear that it maps onto anything in particular. This issue can be adequately addressed with Reichenbach's observation that representations of simultaneity are always based on coordinative definitions, which are ultimately based on *stipulation*. The sensorial present simply *picks out* or *points to* a time (a phase of the circadian cycle or a moment of a represented interval). Further, it is important to notice that the sensorial present has a crucial property that only representations have: it allows for cognitive integration with other sensory-motor representations (like the outputs of the clocks). Actually, it *is* the result of cognitive integration. But a representation has content. The representations of the clocks are about duration. What is the sensorial present's content? This question can be answered by identifying the sensorial present as a kind of mental demonstrative.

Joseph Levine (2010) defines a *mental demonstrative* as a representation that affords direct access to what is demonstrated. It is a particular kind of mental demonstrative, one that Levine calls 'token-demonstrative,' because it does not pick out a type that refers to a token. Rather, it picks out a token directly, without the mediation of a conceptual type. The sensorial present directly picks out a token moment in time, during which cross-modal stimuli are registered as simultaneous.

Linguistic expressions that demonstrate or refer directly have specific properties that are relevant for the characterization of mental demonstratives. Reichenbach (1947) called expressions that play this directly referential role 'token-reflexive.' The reflexivity of these expressions consists in the fact that the utterance *itself* e.g., 'I', 'here' or 'now,' is what directly designates a person, location or moment in time. These expressions are important and unique because they change reference as a function of context, and also because they are not translatable to non-indexical expressions (hence their name, *pure* or *essential* indexicals) (see J. Perry, 1979).[16]

But the point I want to make with respect to the sensorial present is psychological, rather than linguistic. The temporal anchor, or the mental equivalent of the indexical 'now,' must depend on a causal-like, perceptually-driven mechanism that directly refers to moments in time. Just as there are, in any language, expressions such as the pure indexical 'now,' which refer directly to a token moment in time, there must be *representations* that behave in exactly the same way. Among these representations there must be a *temporal mental demonstrative*. I will argue that the sensorial present is the temporal demonstrative for motor control representation and that the calibration mechanism that specifies the sensorial present is a meta-semantic mechanism. 4.2.1. discusses the importance of meta-semantic mechanisms, and 4.2.2. focuses on the temporal meta-semantic mechanism.

4.2.1. *The Need for Meta-Semantic Mechanisms*

As mentioned, something like a temporal mental demonstrative is crucial to anchor the representations of the clocks. In general, mental demonstratives, particularly those equivalent to the so-called 'pure indexicals,' play a fundamental role in theories of mental content, because they anchor or center a subject at a place and time. One way of studying mental demonstratives is through language, by analyzing the way in which indexical expressions refer. Levine (2010), following Kaplan (1989b), distinguishes two kinds of questions that need to be addressed in order to study the content of mental demonstratives (and mental content in general): semantic and meta-semantic questions.

Levine says that semantic questions concern the specification of the content of representations, while meta-semantic questions concern the conditions by virtue of which such representations have the content that they have. Philosophers have written extensively on the semantic question concerning pure indexicals and other demonstrative terms. The meta-semantic issue on which I will focus in this section (i.e., how the temporal indexical *refers* and acquires its content), however, demands a *psychological* explanation. An explanation of the temporal mental

[16] D. Kaplan (1989a) distinguishes pure indexicals e.g., 'I', 'here' and 'now' from demonstratives e.g., 'this' or 'she', because although both vary their reference as a function of context, indexicals only require context to determine their reference. Demonstratives require context *and* ostention or demonstration.

demonstrative must account for the brain's ability to generate a causal-like, immediate link with the environment, such that a temporal referential relation is established.

The sensorial present is the best candidate to fulfill the role of the temporal mental demonstrative. The nervous system, through processes of cognitive integration performed by the calibration mechanism, *points to*, or *picks out*, behaviorally relevant moments in time. The information concerning stimuli that are represented as simultaneous is *reliably* specified on the basis of thresholds for distance and speed. The meta-semantic mechanism that explains why the content of temporal demonstration happens to be the duration of the sensorial present is the calibration mechanism. This is a reliable mechanism that systematically produces representations with information that allow organisms to succeed in navigation. As mentioned, reliability is crucial to give an account of the epistemology of immediately justified beliefs about time. The calibration mechanism satisfies the requirement of being a reliable mechanism, as demonstrated by the findings on the different thresholds involved in the process of integrating the sensorial present.

4.2.2. *The Meta-Semantic Mechanism for Temporal Demonstrative Representation*

I shall now describe in more detail how the temporal mental demonstrative (the semantic characterization of the sensorial present) and the calibration mechanism fit in a broader theoretical framework for mental representation. Levine (2010) distinguishes two kinds of meta-semantic mechanisms: *Intentionally mediated meta-semantic mechanisms*, which depend essentially on the content of other representations to secure their reference, and *direct meta-semantic mechanisms*, which do not depend on other representations to secure their reference.

Direct meta-semantic mechanisms are important to explain how the mind connects with the world and also why certain perceptual representations seem to be *en rapport* (or in an unmediated direct relation) with features of the environment. The calibration mechanism is directly related to the environment because it is constantly and reliably integrating information about speed and distance. Thus, the calibration mechanism is a direct meta-semantic mechanism. One can illustrate this by contrasting the calibration mechanism with the clocks.

Assume that phases and segments of intervals are labels for moments in time in a representation space (similar to a calendar). Focusing on the circadian clock, suppose that an animal phase-locks its circadian clock to the moment in which food is available, say 3:30 a.m. In this case, the circadian clock would work as a temporal intentionally mediated meta-semantic mechanism because it would refer to an individual moment in time, i.e., 3:30 a.m. via another representation concerning the present. Certainly, a phase of the clock *by itself*, say the one that corresponds to 270°, does not refer directly to a moment. Rather, it refers indirectly, in relation to other phases and based on the isomorphism that exists between the cycle of the circadian clock and the period of the rotation of the earth on its own axis. 270° labels a phase that is relevant for behavior in a way that resembles a description, which highlights the need for a direct meta-semantic mechanism that links this mental name-like representation with a *specific* moment in time. In other words, the representations of the clocks need to be *anchored* by the mental equivalent of the indexical 'now.'

'Now' is not translatable to non-indexical expressions. No name, label or description of a moment can capture the content of 'now.' If one calls a moment 'phase 358°' and describes it as 'the moment at which the meeting takes place,' one is not providing any information as to what *particular moment* in time this name and description refer to. Suppose you want to know which particular moment 'phase 358°' refers to, and I give you more coordinates and descriptions (e.g., 'phase 358°' correlates with 12:00 pm, the time called 'noon'). Although you have *more* information about the reference of '358°' you still do not know what *specific* moment it refers to, until you are able to "point at it" somehow and determine that *now* is noon, and thus, that you are now at phase 358°.

You may have temporal representations, like 'phase 358°', or 'today at 12:00 pm.' but these representations cannot be correlated with token moments in time without the existence of a direct link between the sensory-motor system and the environment, which the calibration mechanism provides. Thus, this mechanism is not only responsible for determining cross-modal simultaneity relations. It also plays the important role of being the direct meta-semantic mechanism for sensory-motor temporal representations, particularly clock representations. Actually, the sensorial present might be the most fundamental type of temporal indexical, upon which the linguistic temporal indexical is based.[17]

[17] Linguistically, 'now' varies its duration depending on context and *pragmatics*, e.g., 'Now it is 2012,' 'Now it is time to go to school,' etc. In contrast, the calibration mechanism determines the most fundamental and unmediated indexical reference for

The sensorial present is the first component of the two-phase model of the present that I will outline in the final section of this chapter, section 4.4. In section 4.3., I will describe the second component of the two-phase model, which is the phenomenal present. But before proceeding, I would like to address a possible objection to my characterization of the sensorial present, which is based on one of the typical objections against presentism (the metaphysical thesis that postulates a fundamental and primitive property of 'presentness'). The objection is: compare 'now' with 'here.' It would be odd to say that there is a meta-semantic mechanism for 'here' used by the sensory-motor system. Rather, one can construe 'here' in terms of *limitations* in acuity. If two objects are at an angle in which they overlap in one's visual field, one perceives them as occupying the same region, but it would be odd to say that there is a meta-semantic mechanism at work every time this overlap occurs.

This objection is confused on two fronts. First, maps for sensory-motor navigation need to be anchored to specific regions represented by *different* modalities. Thus, a similar integration process must occur *across modalities*, such that a cross-modal map of regions is directly anchored to the environment (a map of locations that can be used at the organism level, rather than sub-personally). This is the kind of integration process that the calibration mechanism performs in order to represent cross-modal simultaneity. Thus, the objection confuses an issue of the limitations of sense-specific acuity (visual, auditory, etc.) with cross-modal integration processing at the organism level.

Second, there might be many different locations represented at the same time, but not many simultaneities represented at the same location. The calibration mechanism reliably integrates information about distance and speed from different locations, and represents stimuli from such locations as simultaneous. This is why it can be used by the sensory-motor system to refer directly to times. But nothing like this is possible for a single 'here,' unless one has in mind the cross-modal integration of all the locations relevant for action at the organism level, which includes many sense-specific regions, and not a single one. It is possible to point to different regions at the same time, but not to different simultaneous times at a single region. Actually, this is an essential aspect of the *definition* of what simultaneity is (See Reichenbach, 1958; and Geroch, 1978).

moments in time reliably and systematically, based on calculations about speed and distance.

Simultaneity *requires* the selection of a single moment in time for motor control.

4.3. The Phenomenal Present

Can the sensorial present explain how we *experience* the present? An affirmative answer to this question would be: by determining the simultaneity of cross-modal stimuli, the sensorial present anchors sensory-motor representations, providing the sensory-motor system with temporal slices of sensorial information, from which our experiences of change, temporal passage, and motion emerge. The sensorial present explains how we experience the present, because it is responsible for *grounding* experiences.

As mentioned, the cinematic and the retentional models seem to take this route because of their endorsement of the principle of simultaneous awareness. The difference between these models is that the cinematic model does not require the postulation of retentions (i.e., past sensorial memory-like stimuli) that are simultaneously represented with current information in a single sensorial present. The nature of retentions, which are mental representations that are not exactly memories, is controversial, but the main issue is that information is somehow simultaneously presented. As the name suggests, the cinematic model only requires that the "frozen" temporal slices be organized and linked successively, in such a way that they generate the *appearance* of change and motion, in the same way as a set of pictures generates a movie, if appropriately related in a temporal succession. However, this model does not postulate retentions. Current sensorial information is, accordingly, sufficient to explain temporal consciousness.

A negative answer to the question, whether the sensorial present can explain how we experience the present, would point out that what is meant by the term 'experience' is the *conscious* experience of the present, which involves an *experience of succession*, rather than a mere *succession of experiences*. The sensorial present is at best a crude approximation to the experience of succession because it lacks the necessary structure to explain how the mere succession of experiential temporal slices can produce the continuously integrated experience of succession, which includes the experiences of change and motion. The extensional model takes this route, by stating that the specious present (i.e., the briefest moment in which one can experience succession and change) extends beyond temporal slices of simultaneity and includes a set of them.

All the philosophers that work on the topic of temporal consciousness claim that these models are incompatible. As I explained before, I think this may be based on a confusion, which can be eliminated by distinguishing the sensorial present from the phenomenal present, as the two-phase model does. One advantage of the two-phase model is that one need not postulate retentions to explain how slices of simultaneity generate the experience of succession. If there are two different kinds of 'present,' then the extensional model could be compatible with the cinematic model because it would solve a *different problem*, i.e., how to account for our experience of succession and the unity of contiguous conscious experiences.

Thus, the cinematic and retentional models would answer the question of how information about represented *simultaneities* for motor control become available for thought, decision, reporting and action. The extensional model would be exclusively concerned with phenomenal consciousness, while the cinematic and retentional models would be exclusively concerned with access consciousness. But since no retentions are now necessary, because unconscious motor control needs no experiential memory-like representations, the cinematic and extensional models would suffice.

However, some important questions remain unanswered. We seem to have a good understanding of how simultaneity is represented for navigation and motor control, and the evidence provided in this chapter shows that the sensorial present is *specious* in the sense that it has a brief duration. But what is the evidence supporting the thesis that there must be an experiential or phenomenal present? Does it also have a brief duration? How are the contents of the phenomenal present structured, if they are not structured by simultaneity? In the remainder of this chapter, I try to answer these questions. Since the phenomenal present is fundamental to understand temporal consciousness, the evidence for its structure and properties must include introspection.

Using a famous example, suppose you are listening to a song—or even better, put on some music and *listen* to one of your favorite songs. As the notes pass by, you are aware not only of a momentary note at a time (certainly not the time determined by the sensorial present: at most 250 ms). While you listen to the song, you are not listening to a set of *momentary notes* or successive collections of notes at a time. Your experience of the song simply cannot be captured by a sum of auditory sensorial presents because you seem to be listening to *the song*, with all its harmony and unity.

Moreover, a crucial part of your experience of listening to the song is that you expect certain notes to come. The experienced harmony of the song you are listening consists in your retaining the notes that have just passed, unifying them smoothly with those that are occurring now and also with those that you are expecting immediately. Disharmony, or at least surprise, would ensue if you suddenly listened to a completely unexpected set of notes, or if you could not retain the sound of the notes that have just passed. This becomes a lot more interesting and complicated if you are dancing and singing at the same time. You are not experiencing your dancing as a *sum of momentary body movements*. You simply experience it as *dancing*, and more specifically as dancing *to the music: your favorite song*; likewise with singing.

The sensorial present cannot explain your experience of listening to music because of its essential characteristics (i.e., its short duration and its defining relation for stimuli, simultaneity, rather than chronological order, which is the asymmetric relation required for the retention and expectation of notes). What, then, *can* explain your experience of listening to the music? And for *how long* can you experience a stream of sounds as *presently experienced* before they turn into memories? The extensional model seems well suited to explain these characteristics because of its commitment to accounting for the continuity of the experiences you are aware of, which are *consciously interrelated* (rather than simultaneously related) (see Dainton, 2010).

The phenomenal present has a larger number of sources of content than its sensorial counterpart, an important issue that has not been clearly explained in the literature. Since the phenomenal present includes *any* experience, then even non-sensorial information (which is *of no use for navigation and motor control*) is also represented, integrated and unified. Suppose you have the following experiences: looking at a white wall, feeling sad, remembering an event in your childhood, and having a strong feeling that something bad is about to happen. Of all these experiences, only your visual stimuli are being carefully calibrated by the sensorial present. You can *time* things, including how long you have been looking at the wall, because your sensorial present interacts with the clocks for navigation. But this does not explain how your sadness, memories, etc., are integrated and unified, as if they were colors on a canvas: they are part of a *single* experience that you are having.

Another important issue that has not been emphasized in the literature is that the phenomenal present lacks the metric constraints for accurate calibration in navigation that must be satisfied by the sensorial present. This is crucial to understand the significance of the two-phase model of the present. The lack of metric constraints explains why it is so difficult to account for the length of the phenomenal present. Michel Tye (2003) for instance, suggests that everything that is experienced in one stretch of consciousness (say a whole day of wakeful consciousness, or a dream) constitutes a *single* experience. I will explain why Tye is partly right in saying this, but will also suggest that there are temporal limits for the phenomenal present, based on experimental evidence. Section 4.3.1. analyzes the experimental evidence and section 4.3.2. focuses on the relation that integrates experiences phenomenally, which differs significantly from simultaneity.

4.3.1. *The Phenomenal Specious Present*

There are at least two empirical questions concerning the phenomenal present. The first is what mechanism, or set of mechanisms, can explain its properties, and the second is *how long* is it? In this section, I will address the second question, and I will propose an answer to the first one in section 4.4. Evidence confirms that the phenomenal present is also *specious*, in the sense that it has duration, i.e., it occurs during a specific interval of time (which is not surprising given that we never experience durationless instants).

There are two important findings concerning the duration of the phenomenal present. One concerns the minimum temporal threshold required for content to become consciously experienced. The other relevant finding shows that there seems to be a maximum threshold for the phenomenal present. Evidence concerning the minimum threshold of the phenomenal present comes from the research of Benjamin Libet. Although controversial, Libet's findings on conscious awareness have been considered to be valuable experimental evidence on the temporal dynamics of conscious awareness, with significant implications for the scientific study of consciousness.[18] Libet (2004) summarizes his main finding on the aforementioned threshold as follows:

> If you tap your finger on a table, you experience the event as occurring in "real time". That is, you subjectively feel the touch occurring at the same time that your finger makes contact with the table. But our experimental evidence strongly supports a surprising finding that is directly counter to our own intuition and feelings: The brain needs a relatively long period of

[18] See D. Dennett (1991); D. M. Wegner (2002); and R. Penrose (1996).

> appropriate activations, up to about half a second, to elicit awareness of the event! Your conscious experience or awareness of your finger touching the table thus appears only after the brain activations have become adequate to produce awareness (Libet, 2004, 33).

What interests me about Libet's finding is the period of time that, according to him, it takes the brain to produce awareness: up to about *twice the maximum threshold of the sensorial present* (500 ms). Before discussing the importance of Libet's finding for the duration of the phenomenal present, I shall address two issues that need to be clarified in order to understand the nature and scope of this finding.

The first issue concerns the reliability of our conscious experiences as evidence of psychological processes. Libet's result is not the only psychological experiment that challenges our intuitive subjective experience (in this case our awareness of events as occurring in real time). Psychological findings, from change and inattentional blindness to research in motor coordination, have proved that our conscious access to psychological processes may be misleading and unreliable. Thus, Libet's finding confirms what many other psychological experiments have revealed about the shortcomings of conscious awareness, and it should not be considered controversial merely because it challenges our intuitions.

The second issue is more substantial and can be succinctly captured by the question: how should one interpret the experiments on sense-specific simultaneity windows, where subjects *register* the non-simultaneity of stimuli at durations much shorter than 250 ms? If subjects are able to *distinguish* the non-simultaneity of very short events, which presumably requires their *being aware* of such non-simultaneity, what could Libet be talking about when he says that it takes up to 500 ms for *their brain* to elicit awareness of an event? Libet's response to this question is:

> We are talking here about *actual awareness* of a signal, which must be clearly distinguished from the *detection* of a signal. For example, human and nonhuman beings can discriminate between two different frequencies of tactile vibration, even though the intervals between the two pulses in each vibration frequency are only a few milliseconds [...] The ability to detect differences in millisecond intervals is undeniable, but when is one *aware* of that detection? Becoming consciously aware of the difference is what requires the relatively long time. In other words, detection leading to some response can occur unconsciously, without any awareness of the signal (Libet, 2004, 33–34).

We are indeed *unaware* of a lot of sensory-motor information that plays a major role in our daily life (as discussed previously). Yet, Libet's response is unsatisfactory because it relies on the ambiguity of the term 'awareness' without fully clarifying what *he* means by awareness. Libet's lack of clarity, however, does not mean that he lacks experimental evidence. His findings *do* suggest a cerebral delay for awareness, which as Libet highlights, is not at odds with other psychological findings. More importantly, once the distinction between the sensorial present's simultaneity representations for unconscious motor control and the phenomenal present's unified experiences is understood in terms of access and phenomenal consciousness, the confusion dissipates, which is an important advantage of the present proposal.

Libet emphasizes that "cognitive, imaginative, and decision-making processes all can proceed unconsciously, often more creatively than in conscious functions" (2004, 100). Evidence shows that crucial sensory-motor information that is plainly in view and can become consciously available upon *attending to it*, is processed by the brain even though the subjects are unaware of such information.[19] Pylyshyn (2007) calls this 'vision-without-awareness.' The findings on vision-without-awareness include: "change blindness, inattentional blindness, visuomotor control without conscious awareness, blindsight, visual agnosia, and disorders of visual-motor coordination" (Pylyshyn, 2007, 144). As mentioned before, motor control is largely unconscious (see D. A. Rosenbaum, 2002). Any adequate philosophical account of temporal experience must take into consideration these important findings.

Suppose we accept Libet's interpretation and agree that conscious awareness occurs around 500 ms after the event that triggered the relevant neural activation. Is there any other experimental evidence that would explain other properties of the phenomenal present, such as its *upper threshold*? Fortunately there is. But before commenting on this research, I shall state more precisely Libet's proposal for a cerebral delay of "up to 500 ms." In the context of a discussion on the implications of his findings with respect to the issue of free will, Libet suggests that in the case of a subject's awareness of her own will or intention to act, the lower threshold of the phenomenal present, which is up to 500 ms, *may* be shorter: 400 ms. Even in the famous skin stimulation experiment, Libet used a 300 ms stimulus and he frequently characterized the cerebral delay as an interval of 300–500 ms (2004). This is the reason why some authors

19 See also M. Velmans (1991).

say that Libet's results show that there is a delay between readiness potential and awareness of *at least* 300 ms.[20]

I do not want to endorse Libet's controversial claims about free will. What I want to emphasize is that his findings could be interpreted analogously to the findings concerning the sensorial present: they demonstrate the existence of a temporal *threshold* that seems relevant for conscious awareness and is consistent with other psychological findings.[21] Just as the 30–40 ms threshold determines the lower boundary for sense-specific chronological order, the 300–500 ms threshold determines the lower temporal boundary of the phenomenal present.

What is the upper temporal boundary? Research on binocular rivalry and ambiguous or multistable images, such as the Necker cube, provides a plausible answer to this question. Multistable images are *perceptually* interpreted in mutually incompatible ways. For instance, the Necker cube can be perceived as a cube with a lower *or* an upper front but never as *both*. These mutually incompatible interpretations "flip" or alternate spontaneously. Gestalt phenomena, which have always been defined as irreducible to the *summation* of discrete configurations (something that is relevant for the distinction between the experience of succession and a succession of experiences), are also ambiguous, such as the "duck-rabbit" image. These interpretations also alternate spontaneously in the same way as the Necker cube. Psychologists since Wilhem Wundt, have demonstrated that it takes from 2.5 to 3 seconds to *integrate sequential stimuli into experiential units* (for instance a sequence of sounds into a unified rhythm). In the case of ambiguous images, one can attend only to one of the incompatible interpretations, i.e., one can attend to the rabbit or to the duck, but not to both. What is interesting is that these interpretations spontaneously alternate at the *same rate*: every 2.5 to 3 seconds. Pöppel explains,

> Ambiguous figures allow us an interesting insight into the dynamics of the process of consciousness. A content of consciousness can apparently persist

---

[20] See, for instance, Z. W. Pylyshyn (2007).

[21] For example, the distinction between unconscious and conscious information processing can also be explained in terms of action-outcome (A-O) associations based on declarative memory for conscious information, and stimulus-response (S-R) associations based on procedural memory for unconscious information. It is known that after training animals, A-O associations can become S-R associations (See Cheng, et al., 2007). This is actually a quite familiar phenomenon, such as when one learns how to drive: a lot of attention and conscious monitoring is necessary at the beginning, but unnecessary once one *learns* how to drive. The threshold found by Libet could explain the A-O/S-R divide by appealing to the decrease in time involved in the transition from A-O to S-R associations.

> up to circa three seconds. If nothing new is presented requiring other events in the environment to be acknowledged, the alternative perspective thrusts itself automatically into the foreground of consciousness. If still nothing new occurs—if we are again, that is, not "diverted"—then after a few seconds the first perspective returns to consciousness, and so on. After a few seconds, then, the capacity for integration is exhausted (Pöppel, 1988, 60)

The phenomenal present can be characterized as the *window of integration for conscious experiences*. The main role of the sensorial present is to anchor the outputs of the clocks and other sensory-motor representations in general, by providing a cross-modal representation of simultaneity. In contrast, the main role of the phenomenal present is to melt or amalgamate into an experiential whole the information specified by discrete instantiations of the sensorial present with other relevant information, as the extensional model suggests. In the case of ambiguous images, the phenomenal present integrates sequential information into a single *unified experience*: the experience of seeing either a duck or a rabbit.

The same thing happens when there is no perceptual ambiguity: the phenomenal present integrates the information from a set of instantiations of the sensorial present into a phenomenal unit. As Gestalt psychologists have asserted, the phenomenal unit cannot be reduced to the mere sum of discrete representational items, i.e., the *image does not change*; yet it can be *experienced* either as a duck or a rabbit, but never as both. Pöppel calls this "the capacity for integrating chronologically sequential events into a present gestalt" (2004, 63). This seems to be evidence in favor of the distinction between the experience of succession and the mere succession of experiences, because no succession of experiences of the same image suffices to explain how two alternate interpretations of such image are *experienced in succession*.[22]

In sum, the evidence suggests that the phenomenal present has an upper threshold of 3 seconds and a lower threshold of 300–500 ms. Provisionally, the principle of phenomenal unification can be characterized as follows:[23]

---

22 Incidentally, this conscious mechanism that alternates the way in which a single stimulus is experienced seems to be fundamental to explain what S. Siegel (2010) calls 'phenomenal contrasts.'

23 It is important to mention that the evidence concerning the upper temporal bound of the phenomenal present involves a variety of different experiments and seems to be more robust than the one regarding the lower limit (Libet's experiments). See for instance E. Pöppel (2004) and references therein.

> **Phenomenal Unification (Sensorial)**: Once a set of perceptual simultaneities determined by consecutive anchorings of the sensorial present is integrated into a phenomenally unified experience by the phenomenal present, the discreteness of these successive simultaneities disappears and the experience is continuously integrated with other conscious experiences.

Clearly, the relation responsible for phenomenal unification cannot be simultaneity. Notice that it does not help to say that it could be *apparent* simultaneity, because the contents of any experience, say of listening to a portion of a song, do not appear to be simultaneous. Apparent simultaneity seems to be a conceivable conscious experience. The point is that no conscious experience seems to be a sum of successive experiences. Therefore, the lesson *seems to be* that even putative cases of apparent simultaneity *would* involve interrelated experiences that stretch over time. In other words, there would be no *pure* apparent simultaneity at the conscious level, because all conscious experiences involve the experience of succession (i.e., conscious awareness never "freezes").[24]

So what relation could be responsible for *phenomenal unification*? I address this question in the next subsection. But before proceeding, I shall point to two considerations which show that the definition of phenomenal unification just given must be modified. This definition is only provisional because one needs to know how phenomenal unification works, even when it is not in *direct relation to the sensorial present* (when no sensorial information is phenomenally integrated, and one has non-sensorial conscious experiences).

In many cases, the contents of experiences that are unified will be sensorial. But in other cases they will not. This is because the phenomenal present has information from sources that are not strictly sensorial. When one sees colors in dreams, or when one hallucinates, these experiences are phenomenally integrated; yet the sensory-motor system for motor control and navigation is either "off" or not working properly. So there are many occasions in which experiences are phenomenally integrated (e.g., dreams) but the sensorial present is not even operating.

Moreover, what exactly are the consequences of the lack of metric constraints for navigation and motor control with respect to the process of phenomenal integration? In dreams one has phenomenally unified experiences, but the system for metrically calibrated navigation is not

---

[24] See section 4.4. for more on this issue.

operating. This lack of metric constraints, however, does not seem to be an exclusive property of phenomenal unification *in dreams* (what could be called 'oneiric unification'). Rather, this seems to be a robust feature of *any kind of conscious experience*. These considerations should inform the characterization of phenomenal unification.

4.3.2. *The Relation Responsible for Phenomenal Unification*

Dainton (2000, 2004) offers an analysis of the relation responsible for phenomenal unification, which he calls 'co-consciousness,' a term that is also used by previous authors. His analysis includes temporal thresholds and a description of co-consciousness as a primitive relation. In one of the relevant passages in which Dainton discusses synchronic and diachronic co-consciousness, as well as their relation to the 'specious present,' he says:

> I will use the (less than optimal) term 'specious present' to refer to those brief phases of our streams of consciousness during which we are directly aware of change and persistence. Since a specious present has some (apparent) temporal depth, it has (seemingly) earlier and later parts; and since the transition between these parts is directly experienced, the parts are *phenomenally connected*—they are co-conscious, but diachronically rather than synchronically (Dainton, 2004, 374).

Dainton says that the term 'specious present' refers to brief phases of our stream of consciousness that can be connected synchronically and diachronically. How long are these brief phases? He quotes Pöppel's evidence concerning the order threshold and says that the specious present's lower threshold is 30 ms (2004, 374). This duration is somewhat compatible with my characterization of the sensorial present, but it is incompatible with my proposal for the phenomenal present. More importantly, the evidence suggests that this threshold is only relevant for sensorial motor control.

Another problem arises from what Dainton says about the upper bound of the phenomenal present. He defends an overlap model for the connection between specious presents, according to which "a *stream* of consciousness does not consist of a succession of self-contained chunks or pulses of experience, laid end to end like a row of bricks." Rather, Dainton continues, "We must recognize the phenomenal connections *between* them" (2004, 378). He realizes that in order for the overlap model to work, there must be an upper bound that temporally delineates each specious present. Dainton *mentions* the 3-second upper limit (which is the upper limit that I propose for the phenomenal present) and then writes,

> The figure of three seconds [...] is based on people's ability to discern distinctive, memorable, or pleasing patterns in their experience, *temporal gestalts*: think of how the notes in a musical phrase, or the words in a line of spoken poetry hang together, or seemingly form natural units. However, given that these patterns extend quite some way through time, there is no guarantee that the beginning and end of a given pattern fall within the scope of immediate experience. For my own part, I would tentatively estimate the duration of my typical specious present to be half a second or less (Dainton, 2000, 171).[25]

There are three problems with Dainton's characterization of the upper bound of the phenomenal present. First, why would Dainton opt for an introspective estimation? Second, the experimental evidence has nothing to do with "patterns." Rather, such evidence has revealed, as Pöppel says, the temporal dynamics of *awareness in general*. Think about the experiment with ambiguous images. What is the pattern in this experiment? The image *does not change at all*, so there is no pattern in the stimulus. One of the perceptual interpretations comes to the foreground and the other one recedes, and they keep flipping every 3 seconds even though the image is the same. Third, which is the most substantial problem, why should the upper bound be 500 ms or *less*? This value is what Libet specifies as the *minimum* threshold required for sensory-motor information to become conscious.

But these issues could be ignored if what one wants is an account of the relation of co-consciousness. What *is* co-consciousness? Dainton characterizes co-consciousness as a primitive relation: a feature of any conscious experience, analogous to color or sound, unexplainable and non-analyzable (2000, 236). He also says that it is *transitive* at "any given moment shorter than the specious present," but that co-consciousness over time is *not transitive* (2000, 172). This may sound contradictory, but it merits further explanation. Dainton provides the following example: three total experiences (which I am about to define) X, Y and Z "can be such that X is co-conscious with Y, Y with Z, but X is not co-conscious with Z" (2000, 172). A total experience (a unified set of experiences at a time) includes not only perceptual and other sensory-based experiences, but also experiences associated with thoughts, feelings and emotions. Dainton claims that co-consciousness unifies all these experiences as follows: "Co-consciousness is a relation, its terms are experiences, and so far as individual total experiences are concerned, it is reflexive, symmetrical

[25] Dainton (2010) now suggests it is around 750 ms, which is also problematic.

and transitive" (Dainton, 2000, 219). Dainton's characterization of co-consciousness explains how unity and a degree of continuity are achieved by the symmetrical and reflexive linking of all experiences at a "moment in time." But how should one interpret the *non-transitivity* of diachronic co-consciousness? As Dainton acknowledges, the non-transitivity of co-consciousness in the synchronic case is *incomprehensible*, but according to him, it seems to be *inevitable* in the diachronic case:

> To the extent that our streams of consciousness require non-transitive co-consciousness, it is tempting to suppose that non-transitivity and temporality are essentially linked, and this further strengthens the case for thinking that synchronic co-consciousness cannot fail to be transitive (Dainton, 2000, 182).

Why is it that synchronic co-consciousness cannot fail to be transitive? Presumably, this has nothing to do with simultaneity. But then a synchronous non-simultaneity seems like a very hard pill to swallow. One might also ask why diachronic co-consciousness is linked with temporality, but not the synchronous version, unless synchronous co-consciousness is just simultaneity? The problem is that Dainton defends an extensional model, which is incompatible with the models that assume simultaneity as the main relation among conscious experiences (the cinematic and retentional models). Synchronous co-consciousness cannot, therefore, be simultaneity. Moreover, simultaneity is always transitive, unlike co-consciousness, which can be non-transitive. Synchronous and diachronic co-consciousness must be the same relation (the relation responsible for phenomenal unification). But how can the same relation be transitive in some cases and non-transitive in other cases?

Whatever co-consciousness is, it is an odd kind of relation. Maybe the best way to understand it is in terms of a phenomenal form of *binding* that comes in degrees, similarly to a resemblance relation. A particular shade of red will look the same as any other shade that is very close to it in the continuum. At some point, though, this will weaken, and the relation 'looks the same as' will no longer be transitive with respect to those shades of red that are far away in the continuum.[26] So, maybe co-consciousness unifies experiences in terms of the degree of *salience* of information, which depends on our limited capacity for attention, and it breaks down in an analogous way to the resemblance relation (which would explain diachronic co-consciousness). One can illustrate this as follows.

[26] This is what generates the so-called 'phenomenal sorites.'

Suppose that you are angry because of a speech you are watching on television. Co-consciousness unifies your experiences of anger (maybe combined with frustration) with visual and auditory experiences of a specific person and a specific voice. It is very likely that you are unconsciously registering other sensory-motor information, and a significant amount of this information might potentially become part of a unified experience. But at the moment in which you are angry listening and watching the speech, this information is simply not part of what you are experiencing (it is, as it were, in the background). Now suppose that while you are angrily listening to the speech, you are eating something. You swallow what you are eating and then focus your attention on what you just swallowed and ask: what am I eating? While you were angry, you were not noticing the flavor of what you were eating, but this does not mean that you were not processing this information. Once you focus your attention on the taste of what you were eating, you realize it is an apple, which you confirm visually by looking at the apple.

This story seems plausible, but unfortunately, attention is doing most of the explanatory work, and one needs more details about how co-consciousness (of the synchronic and diachronic varieties) relate to attention. I will say more about this in section 4.4. For now, it seems safe to say the following, which remains quite abstract, about the co-consciousness relation. Co-consciousness does not *entail* transitivity or intransitivity, i.e., it is a *non*-transitive relation. Co-consciousness does entail symmetry and reflexivity. This explains, for instance, how one can have a fringe feeling of expectation, and how this feeling may become co-conscious with the hope of seeing a specific person. It seems that co-consciousness satisfies the following principles:

> **Symmetry**: If an experience X is co-conscious with another experience Y, then Y is also co-conscious with X.
> **Reflexivity**: Any experience X is co-conscious with itself.
> **Non-transitivity**: If an experience X is co-conscious with another experience Y, and Y is co-conscious with another experience Z, then it does not follow that either X is co-conscious with Z or that X is *not* co-conscious with Z.

I think this is the most plausible way of understanding Dainton's proposal for synchronic and diachronic co-consciousness. However, this is hardly an adequate account of co-consciousness (many relations may have these abstract properties), and it is, at best, the very beginning of one because very intricate issues remain unsolved. In particular, if there is a parthood relation between total experiences and their sensorial and non-sensorial experiential components (which seems crucial to explain the unity of consciousness), then it is important to understand the metaphysical underpinnings regarding parthood and co-consciousness.[27]

For instance, 'being part of' seems to be transitive within one total experience, but two *total experiences* that are co-conscious cannot, by definition, be part of each other ('being part of' is anti-symmetric). So it seems hard to square a transitive and anti-symmetric relation with a non-transitive and symmetric one. This is such a central problem for the extensional model that Dainton (2010) recently suggested that consciousness might have the structure of *atomless gunk*, as a potential reply to the parthood objection. But then, a theory of conscious atomless gunk must be provided, and I am afraid there are no specific forthcoming details about that theory.

For related reasons, Philippe Chuard (2011) argues that there seems to be no good candidate for a relation of diachronic co-consciousness. I agree with Chuard that parthood, or similar candidates, do not seem adequate to fulfill the role of co-consciousness. However, I disagree with his conclusion that this is because *extensional models* are inadequate. Thus, I agree with Dainton and other proponents of the extensional model that there is a distinction between the experience of succession and a succession of experiences. The failure to characterize one component of a model does not entail the failure of the model and its basic assumptions. So, instead of giving a mereological characterization of the allegedly primitive co-conscious relation for experiences (what unifies them), I shall describe the general constraints that any candidate proposed to fulfill the role of co-consciousness must satisfy.

I argued that simultaneity is the relation that unifies stimuli in the sensorial present, and that a similar unifying relation is required to account for the interrelatedness of experiences in the phenomenal present (something like co-consciousness). As I briefly mentioned, two aspects of phenomenal unification that deserve more theoretical attention than they have received so far are the multiple sources and varieties of conscious awareness that can be phenomenally unified, on the one hand, and the lack of metric navigational constraints for phenomenal unification,

[27] See T. Bayne (2010), particularly chapter 2, for a mereological account of phenomenal unity that construes the subsumption of experiences at a time by a total experience in terms of parthood. Incidentally, Bayne assumes the principle of simultaneous awareness.

on the other. I shall elaborate on these aspects before proposing a revised version of phenomenal unification.

Cases of phenomenal unification of multiple experiential sources, sensorial and non-sensorial, are well known in the literature. A total experience of all your sensorial input (e.g., visual, auditory, haptic, proprioceptive, etc.) is unified with your emotions, thoughts, plans, and expectations as you, for instance, walk down the street. You do not have to be attending to all of them (attention seems to be what makes information *accessible* to you), for them to be unified in your total conscious experience of walking down the street.

The cases that have not received enough attention in the literature are cases involving the phenomenal unification of different forms of conscious awareness. Dreams are particularly relevant to illustrate these cases. Suppose you have a vivid dream of going to a park and seeing a friend. You then wake up, remember the dream, and decide to call your friend. One may say that the first form of consciousness (dreaming) is strictly non-sensorial, because the sensory-motor system is literally inactive, while the second form is sensorial (wakeful consciousness). There is something it is like to wake up from a vivid dream (radically different from *just waking up*) and acting upon something that happened in the dream. This phenomenal unification involves not only a variety of sources of information, but also two distinct types of conscious awareness. Any account of phenomenal unification should explicitly address this issue.

One objection to this type of unification is that there are transitions that sever the unity of these experiences. This is true, but only if one is not dreaming vividly right before one wakes up. In the case of a vivid nightmare, the experience of fear remains unified with the non-oneiric experiences that one has immediately after waking up. Memories of the dream are also unified with other memories, and being aware of these memories is also subsumed by a total experience. Thus, the amount of information integrated by phenomenal unification can be highly diverse.

The other important constraint on an account of phenomenal unification concerns the irrelevance of metric-navigational information for the integration of total experiences. This is a tricky issue. Although the phenomenal present has an upper threshold for integrating total experiences, phenomenal unification may relate these total experiences for much longer durations than this upper threshold. This clarifies what I said before (in section 4.3.), when I suggested that Tye (2003) is partly right in saying that whole stretches of consciousness may be a single conscious experience. Although there are temporal limitations with respect to the upper threshold of the phenomenal present (part of the temporal dynamics of conscious awareness) this threshold seems to be irrelevant with respect to how many, and for how long, total experiences may be phenomenally unified. The same holds for any metric information concerning navigation, including sensorial simultaneity: it is irrelevant for phenomenal unification.[28] Taking these two constraints into account, one can define phenomenal unification as follows:

> **Phenomenal Unification (General)**: Once a set of experiences is unified into a *total experience*, regardless of whether these experiences are sensorial or non-sensorial, the total experience appears subjectively as a single conscious episode, continuously interrelated with other total experiences. This holds, necessarily, for all forms of conscious awareness (including those that are strictly non-sensorial), regardless of any metric or psychophysical thresholds.

This is still a very general characterization of phenomenal unification, but it has the merit of not endorsing controversial views about experiences, while incorporating important insights from the empirical evidence and the literature on conscious awareness. An advantage of the two-phase model is that it *explains* the irrelevance of metric information for phenomenal unification (motor control occurs unconsciously). Another advantage of this model is that it explicitly assumes that there are more sources of phenomenal unity than the sensorial present can provide. In this way, the two-phase model seems to accommodate all the scientific evidence on motor control and conscious awareness. However, my proposal for a two-phase model of the present is incomplete without addressing the question: how do the sensorial and phenomenal present relate to each other? I answer this question in the next section.

4.4. The Two-Phase Model of the Present

In this section, I outline a two-phase model of the present. The most important aspect of this model is that it explains the relationship between the sensorial and phenomenal present in terms of *cross-modal attention for action selection*. Although attention has been intensely studied in psychology, it has not received systematic treatment in philosophy until

---

[28] The upper bound of the phenomenal present may be relevant for the experience of *presence*, without delineating or dissecting total experiences. I say more about this in section 4.4.

recently, when philosophers started exploring the relation between consciousness (access and phenomenal) and attention. 'Attention' is an umbrella term for different perceptual processes, some of which happen under the guidance of the subject, while others occur automatically and independently of voluntary action. The consensus, however, is that attention is important to make information *phenomenally salient* and *accessible* for different cognitive tasks. Further, the consensus is that attention *anchors* cognitive processes. Thus, attention seems to be a relevant mechanism for the interaction between phenomenal and access consciousness.

Attention, like co-consciousness, can be defined as diachronic or synchronic (see W. Wu, 2011). One can shift attention from one conversation to another, from one visual stimulus to another, etc. In these cases, one is shifting attention from one perceptual stimulus to another, but the stimuli are perceptually available at the same time, synchronously. In contrast, a stimulus may grab our attention suddenly, thereby shifting our attention from what we were attending to before to the stimulus that is now salient. Thus, it seems that synchronic attention is associated with voluntary attention, while diachronic attention is associated with automatic, involuntary attention.

Philosophers have proposed different ways to explain the relationship between the two types of consciousness (access and phenomenal) and these two types of attention. Some argue for identity, stating that phenomenal consciousness is nothing other than synchronic and diachronic attention, or alternatively, that access consciousness is merely attention plus global accessibility. Others say that there is something epistemically unique about conscious attention. The space of possibilities here is considerable. One may say that phenomenal consciousness is not reducible to any form of attention, and that attention is limited in that it operates on phenomenally conscious contents, in order to make them accessible (access consciousness).

As with the models of temporal consciousness, these all seem to be mutually incompatible views. But they need not be, if one adopts the two-phase model of the present. Because of the findings on unconscious motor control, it actually seems that one *must* adopt the two-phase model of the present if one wants to have a scientifically plausible account of conscious attention. Take, for instance, the unconscious anchoring of information from the clocks to the sensorial present. These are representations at the organism level that are indispensible for motor control and navigation. The process is strictly sensorial (i.e., information from the senses is calibrated and interpreted temporally, using information from the clocks). The activation of the stopwatch, for example, seems to depend on sensorial attention.

It would be wrong to say that these navigational processes *require* phenomenal consciousness, because it would flatly contradict the findings on motor control and place unacceptably high constraints on animal navigation. On the other hand, it would be wrong to call these processes sub-personal, strictly automatic, or purely mechanized. Information from the clocks is integrated in navigational processes, available for action, decision-making, and other representational processes. The integration and anchoring of this information is dependent on a cross-modal representation of simultaneity at the organism level: the sensorial present. Since this information becomes available for the organism as a whole, it is adequate to say that the sensorial present is associated with access consciousness.

Let us again consider the example of Anna, the baseball player. Anna's conscious experiences about the home run, her emotions, the sounds in the stadium, etc., are smoothly connected, organized in terms of familiar objects, and constantly relevant for action selection. She can choose to raise her hand and wave it in celebration, saluting the crowd, and she will certainly be aware of this. Or she may just have the *inclination* to do so and then refrain from action, thinking that it would be too arrogant. She will also be aware of her inclination to raise her hand, even if no overt action ensues. Having inclinations for action is critical for *conscious action selection*, but action selection can be dissociated from *motor control*. There are two types of integration happening while Anna has this inclination: the motor control integration that depends on fine metric information for navigation, as she paces around the field, and the conscious integration of her experiences, which are the basis of her inclinations. The latter is independent of any metric constraint and occurs within the phenomenal present. The former depends on the sensorial present.

There are several ways to illustrate the dissociation between action selection and motor control. For instance, in the Müller-Lyer illusion, although the subjects' conscious self-report is inaccurate and reflects the illusion's cognitive influence, their motor control (manual behavior) is accurate and *not influenced* by the illusion. This finding seems to suggest that conscious perception does not necessarily have a *direct* cognitive influence on action. However, Stottinger and Perner (2006) have shown that although motor control is not influenced by the illusion, action selection, just like conscious perception, *is influenced by the illusion*. In their experiment, Stottinger and Perner presented subjects with vertical lines grouped in two sets (one with open brackets and the other with closed brackets, as in the standard Müller-Lyer illusion). When asked 'which gang of lines would you fight?' subjects chose the "smaller" lines, although their motor control in the absence of this question did not distinguish between the sets of lines, because it was not influenced by the illusion. This finding demonstrates the dissociation between action selection and motor control. As Morsella and Bargh say, "Illusions are based on inborn or learned knowledge of the ventral stream. This knowledge constrains action selection but not motor control" (Morsella and Bargh, 2010, 7).

The role of accuracy conditions in perception, including those concerning reliable time measurements and representations of simultaneity (e.g., the outputs of the clocks and the sensorial present), is to make successful motor control possible. Notice that subjects successfully objectify and *perceive* the length of the lines, as is manifest in their manual behavior. But this is *unconscious* perception. In contrast, conscious perception is influenced by the illusion and frames action selection accordingly. It is not, therefore, erroneous to say that subjects successfully perceive (unconsciously) and misperceive or misrepresent (consciously) the same stimuli (or external objects). Still, there are other, easier and more intuitive ways to illustrate this dissociation. When Anna dreams that she hits a home run, her awareness presents her with similar inclinations for action selection, but her motor control system is off (she lies in bed as she has these conscious inclinations).

What is crucial for the proposal of the two-phase model of the present is that conscious attention seems to depend on the interaction of *both* the sensorial and phenomenal present. This is an important issue that needs further study and exploration, and I can give only the general outlines of this model. However, the *evidence* and the current philosophical understanding of these issues suggest that the type of metrically constrained anchoring of the outputs of the clocks can only be produced by an unconscious representation of simultaneity that frames motor control and navigation (i.e., the sensorial present). Admittedly, the type of continuity and integrity of total experiences can only be guaranteed by a non-navigational, non-metrically constrained integration process that interrelates sensorial and non-sensorial experiences (i.e., the phenomenal present).

Conscious attention, however, seems to depend on both, because the type of anchoring it provides is *epistemic*: is not strictly navigational. It provides a link between action selection and motor control, and when things go well (when one is not under the influence of an illusion), conscious attention provides epistemic access to navigationally relevant representations. At the same time, what is salient becomes an anchor for inferences, thoughts, decision-making (i.e., access consciousness), and it is also integrated with other experiences into a unified total experience (i.e., phenomenal consciousness). Access consciousness *influences* diachronic attention because stimuli grab it involuntarily. Phenomenal consciousness seems to be necessary for diachronic and synchronic attention and awareness, but diachronic attention seems to be more deeply associated with motor control, which suggests that it is not likely to be present in dreams, for example. But all this needs more study and experimental evidence, remaining speculative at this point.

An objection to this model is: why call the sensorial present 'present' at all? If we are not aware of it, then it cannot be what *we* call the present. There are two replies to this objection. The first is that what we call the present may just be part of the full explanation of what the present is (according to the psychological evidence), just as what we call color is just part of the full explanation of what color is. There is something *it is like to be* aware of present experiences, but that something does not provide a complete explanation of what the present is: the motor control representation of cross-modal simultaneity must also be taken into account. This cross-modal representation is so important that two of the models for temporal consciousness (i.e., the cinematic and retentional models) assume simultaneity. The second reply is that it is perfectly adequate to call the sensorial present 'present' because, as this chapter has shown, the indexing of current information resembles the indexical 'now.' It is, after all, current, or present information that is being indexed.

Another question is: why use the word 'phase' to characterize this two-fold model of the present? My reply to this question is that the term 'phase' is only meant to capture the radical and abrupt change in structure that cognitive representations undergo when they are integrated by the phenomenal present. Information changes in the sense that simultaneity is no longer the main relation unifying information. Abrupt changes in structure, as when a liquid turns into a solid, are called 'phase transitions' in physics. The term 'phase' is supposed to capture this kind of change, from sensorial present information to total experiences unified by the phenomenal present. But this is just a name, and nothing theoretically deep hangs on it.

A third objection is: why assume that attention is not identical with phenomenal consciousness? Isn't this a form of theoretical bias assumed by this model? The responses to these questions are quite straightforward. One response is that this is more than just a theoretical bias: I have been supporting my proposals with scientific evidence, in which the distinction between phenomenal consciousness and the two forms of attention required for access consciousness seems to have been demonstrated, thus lending weight to the two-phase model. There may be room for disagreement regarding my interpretation of conscious attention. But the two-phase model has enough merits, from a theoretical and empirical point of view, to be considered as a plausible candidate for an explanation of the findings.

Another reason not to equate attention with phenomenal consciousness has to do with a philosophical bias. Philosophers tend to focus exclusively on perceptual consciousness in their debates. But there are other forms of consciousness. One that is of enormous scientific interest is what I have been calling oneiric awareness (awareness in dreams). Dreams are interesting because the motor control system, with all its metric constraints, is inactive. Yet, one has conscious inclinations, memories, non-sensorial and dream-"sensorial" experiences. One can experience anxiety and vivid colors in dreams. These experiences are integrated into total experiences that one has while dreaming. But is one *attending* to stimuli? Not in any clear sense of 'attending to.'

One may also have qualms about the temporal thresholds I discussed in this chapter. It is important to emphasize that the two-phase model of the present does not strictly depend on the values I have been relying upon in describing the temporal boundaries of the sensorial and phenomenal present. Even if science finds slightly different values, which is not very likely, the two-phase model would still be adequate: the evidence in favor of the distinction between the sensorial and phenomenal present seems too robust to be false or superficial, allowing one to predict that some values demarcating the two types of present will remain in place.

Finally, it seems odd that the lower threshold of the phenomenal present—the time it takes for information to become phenomenally conscious—is so significant, while the upper threshold seems to be completely irrelevant, because of the nature of the relation responsible for phenomenal unification. This is indeed an intricate problem, which involves the set of difficulties I described previously regarding parthood. However, my proposal is that the relation for phenomenal unification should be construed independently of temporal thresholds. But what, then, is the role of the upper threshold?

Here is a suggestion. Perhaps the upper threshold demarcates phenomenal contrasts that are not based on either attention or changes in the stimuli. This would mean that it demarcates phenomenal contrasts that are entirely based on the dynamics of conscious awareness: *now the duck is present, now the rabbit*. This could be understood as an experience of *presence*, which is integrated with other experiences by phenomenal unification. Phenomenal unification should not for that reason be constrained by such threshold. Nevertheless, this is a topic that requires a significant amount of theoretical and experimental work.

In particular, a remarkably difficult issue is how to construe the *experience* of simultaneity, or apparent simultaneity, which must be integrated with other experiences by means of co-consciousness. This is the source of many puzzles regarding temporal consciousness. One option is simply to deny that such experience is possible (necessarily, any experience involves the experience of succession). Another option is to explain it in terms of co-consciousness, roughly as follows.

A single sound experienced in isolation is integrated according to the thresholds described previously. When that sound is experienced as being simultaneous with another sound, the qualitative character of the experience is different, and the two sounds are subsumed by a total experience. This phenomenal contrast can be understood in terms of *three* experiences: the experiences that correspond to the two sounds and the experience of simultaneity. The fact that the experience of simultaneity is integrated by co-consciousness can only be denied on pain of contradiction (i.e., it would be an experience that cannot be integrated by co-consciousness). This proposal, however, seems to suggest that there is a distinction between synchronous and diachronic co-consciousness, and thus it has the shortcomings that were described above. The problems that this topic generates are substantial, therefore, and need to be at the center of future research on temporal consciousness.

To sum up, the most important advantages of the two-phase model are: a) it is based on scientific evidence; b) it offers a detailed characterization of its components in terms of cognitive function; c) it provides a novel way of addressing philosophical issues; and d) it spells out clearly how the sensorial present relates to the other main temporal mechanisms of the sensory-motor system: the clocks.

Since this model is based on scientific evidence, it is always possible that future research may disconfirm it. But even if the two-phase model proves to be inadequate, it would at least have presented the significant challenge of explaining the evidence it purports to account for with a better model that could accommodate sensorial and phenomenal information processing with the same theoretical plausibility.

One, final question remains unanswered, however: what is the cognitive interaction between the clocks and the phenomenal present, if any? In the conclusion, I suggest that this interaction is best understood as the experience of the flow of time.

CONCLUSION

I have argued that the metric outputs of the clocks determine representation spaces of possibilities that structure and organize the behavior of an organism, by guiding it successfully in the environment and by helping it decide on the best course of action. These representations form reliably produced (perceptually driven) beliefs about duration. I also explained that there are two types of present and provided a *two-phase model* of the present that seems to satisfy empirical and theoretical requirements. I shall conclude with a brief discussion regarding the experience of the flow of time and suggest other potential implications of the proposals defended in chapters 3 and 4.

The topic of the passage of time has been central in the debate on the metaphysics of time. B-theorists say that the passage or flow of time is an illusion, as I noted in the Introduction. Thus, the perception of change and motion are also illusions (nothing in spacetime is changing, moving or passing). In contrast, A-theorists say that the passage of time is an essential feature of spacetime. Change and motion, as well as the pure passage of time, can be accurately perceived because they are objective.

The metaphysics of time seems to be stuck in the debate between A and B theorists, with the B theorists in the majority and leading with allegedly better arguments, based on current physics. However this debate develops, it is clear that physics will have the last word. Without choosing sides between A and B theorists, I would like to make one final remark about the *experience* of the flow of time. This *experience* has been a crucial point of defense for some metaphysicians. Regardless of the structure and nature of spacetime, therefore, what produces the experience of the passage of time, independently of whether or not it is illusory?

One answer is to appeal to integration mechanisms, such as those involved in the integration of the sensorial present. From a set of frozen time slices, integration mechanisms produce the illusion of passage and motion, as assumed by the cinematic and retentional models (see for instance L. Paul, 2010). Yet, this does not really explain why we have the experience of *passage*, if it is understood as the experience of continuous succession (rather than a mere succession of experiences). What could explain the experience of passage? I suggest that the experience of passage is produced by the phenomenal present's interaction with the clocks. I do not have a full-fledged account to offer about this issue, but will propose one possibility.

The phenomenal present has no metric constraints, but it seems that the experience of passage must be linked with mechanisms for representing *duration*, and the ideal candidates are the clocks. The clocks are part of the sensory-motor system, and their representations are anchored by the sensorial present to create the temporal framework for navigation and motor control. As I explained in section 4.4., the sensorial present also interacts with the phenomenal present by means of conscious attention. In this way, there can be an experience of pure passage or duration. Since the phenomenal present is always associated with the integration of total experiences, the experience of passage may depend on sensorial or non-sensorial information (for instance, the stopwatch is activated by attention-related sensorial processes, while the circadian clock is always running).

One can capture this idea in terms of supervenience. The experience of passage supervenes on the outputs of the clocks. This means that the qualitative character of our conscious awareness of the passage of time must somehow exploit the analog continuous nature of these outputs. In other words, there cannot be changes in the continuity of our experience of the passage of time without changes in the continuity of the outputs of the clocks for motor control. Notice that this continuous or gap-free aspect of the experience of the passage of time does not entail the integration of total *experiences* in a stream of consciousness. For this type of integration, one needs phenomenal unification (of the general kind).

The experience of the passage of time is one of the points of contention in the metaphysics of time with respect to which the theses of this book may have interesting implications. Another important matter of debate is our immediate acquaintance with the present and the uniqueness of the present. Since there are two senses of 'present' according to the model I favor, this model may suggest different ways of understanding the claims made by A and B theorists, because metric considerations about the present exclusively concern the sensorial present, while experiential considerations exclusively concern the phenomenal present.

One of the main goals of this book is to provide a unified theoretical account of the findings in psychology that at the same time addresses the most important philosophical problems concerning time representation, the epistemology of basic beliefs about time, and temporal consciousness. There have been previous accounts of psychological findings that are relevant for philosophical problems, but none have focused *exclusively* on the most fundamental representations for time in navigation and basic phenomenal experiences.

The proposals of this book relied on the philosophy of science, the basic requirements for time measurements, the reliabilist view of epistemic justification and empirical evidence in psychology. It explains time representation for navigation without appealing to the self, memories, conceptualizations, causation or any other philosophical terms that introduce complexities that are alien to the topic. What one needs to successfully and reliably measure or represent time are clocks and simultaneity judgments. This is the lesson that physicists learned in their investigations of spacetime and it is a central topic of this book.

The book follows the same strategy in order to give an account of the present moment, which is consistent with philosophical naturalism. Any model of the present must take into consideration the findings on simultaneity windows and the time frame for unified experiences. It seems that the only plausible model, given these constraints, is the two-phase model of the present.

BIBLIOGRAPHY

Aristotle. 1980. *Physics*. Grinnell, Iowa: The Peripatetic Press.

Aschoff, J. 1965. Circadian Rhythms in Man. *Science*, 148: 1427–1432.

Augustine. 1961. *Confessions*. Harmondsworth, UK: Penguin.

Barbour J. 2000. *The End of Time*. New York, NY: Oxford University Press.

Bayne, T. 2010. *The Unity of Consciousness*. New York, NY: Oxford University Press.

Bermúdez, J. L. 1995. Nonconceptual Content: From Perceptual Experience to Subpersonal Computational States. *Mind and Language*, 10(4): 333–369.

—— 2003. *Thinking without Words*. New York, NY: Oxford University Press.

Bhattacharjee, Y. 2006. A Timely Debate About the Brain. *Science* 3–311(5761): 596–598.

Blachowicz, J. 1997. Analog Representation Beyond Mental Imagery. *The Journal of Philosophy*, 94(2): 55–84.

Block, N. 1995. On a confusion about the function of consciousness. *Behavioral and Brain Sciences*, 18: 227–47.

—— 2003. Consciousness. In L. Nadel (ed.) *Encyclopedia of Cognitive Science*. New York, NY: Nature Publishing Group.

Broad, C. D. 1923. *Scientific Thought*. Bristol, UK: Thoemmes Press/Routledge; Reprinted 1993.

Brouwer L. E. J. 1907. On the Foundations of Mathematics. In *Collected Works*. Amsterdam, The Netherlands: North-Holland Publishing Company; 1975; Vol. 1, pp. 13–52.

Burge, T. 2010. *Origins of Objectivity*. New York, NY: Oxford University Press.

Byrne, A. 2009. Experience and Content. *Philosophical Quarterly*, 59, 429–451.

Callender, C. 2008a. The Common Now. *Philosophical Issues*, 18: 339–361.

—— 2008b. Review of Robin Le Poidevin's *Images of Time. Notre Dame Philosophical Reviews*.

Campbell, J. 1995. *Past, Space, and Self*. Cambridge, MA: MIT Press.

Campbell, N. R. 1928. *An account of the principles of measurement and calculation*. London, UK: Longmans, Green.

Carnap, R. 1967. *The Logical Structure of the World*. Berkeley, CA: University of California Press.

Cheng, K. and Roberts, W. 1991. Three psychophysical principles of timing in pigeons. *Learning and Motivation*, 22: 112–128.

Cheng, R. K., MacDonald, C. J., & Meck, W. H. 2006. Differential effects of cocaine and ketamine on time estimation: Implications for neurobiological models of interval timing. *Pharmacology Biochemistry and Behavior*, 85: 114–122.

Cheng, R. K., Hakak, O. L. and Meck, W. H. 2007. Habit formation and the loss of control of an internal clock: inverse relationship between the level of baseline training and the clock-speed enhancing effects of methamphetamine. *Psychopharmacology*, 193: 351–362.

Chisholm, R. 1990. Referring to Things That No Longer Exist. In J. Tomberlin (ed.), *Philosophical Perspectives 4* (*Action Theory and Philosophy of Mind*). Atascadero, CA: Ridgeview, 545–556.

Chuard, P. 2011. Temporal Experiences and Their Parts. *Philosophers Imprint*, 11(11): 1–28.

Church, R. M. and Gibbon, J. 1982. Temporal generalization. *Journal of Experimental Psychology: Animal Behavior Processes*, 8: 165–186.

Church, R. M., Meck, W. H., and Gibbon, J. 1994. Application of scalar timing theory to individual trials. *Journal of Experimental Psychology: Animal Behavior Processes*, 20: 135–155.

Clark, A. 2000. *A Theory of Sentience*. New York, NY: Oxford University Press.

Clayton, N. S., & Dickinson, A. 1998. Episodic-like memory during cache recovery by Scrub-jays. *Nature*, 395: 272–274.

Clayton, N. S., Yu, K. S., and Dickinson, A. 2003a. Interacting cache memories: Evidence of flexible memory use by scrub jays. *Journal of Experimental Psychology: Animal Behavior Processes*, 29: 14–22.

Clayton, N. S., Bussey, T. J., and Dickinson, A. 2003b. Can animals recall the past and plan for the future? *Nature Reviews Neurosciences*, 4: 685–691.

Clayton, N. S., Emery, N. J., and Dickinson, A. 2006. The prospective cognition of food caching and recovery by western scrub-jays (*Aphelocoma californica*). *Comparative Cognition & Behavior Reviews*, 1(1): 1–11.

Concannon, P. W., Parks, J. E., Roberts, P. J., Tennant B. C. 1992. Persistent free-running circannual reproductive cycles during prolonged exposure to a constant 12L:12D photoperiod in laboratory woodchucks (*Marmota monax*). *Laboratory Animal Science*, 42: 382–391.

Cordes, S., Gelman, R., Gallistel, C. R. and Whalen, J. 2001. Variability signatures distinguish verbal from nonverbal counting for both large and small numbers *Psychonomic Bulletin & Review*, 8: 698–707.

Coull, J. T., Vidal, F., Nazarian, B., Macar, F. 2004. Functional anatomy of the attentional modulation of time estimation. *Science* 303: 1506–1508.

Cummins, R. 1996. Systematicity. *The Journal of Philosophy*, 93: 591–614.

Cummins, R., Blackmon, J., Byrd, D., Poirier, P., Roth, M. and Schwarz, G. 2001. Systematicity and the Cognition of Structured Domains. *Journal of Philosophy*, 98: 167–185.

Dainton, B. 2000. *Stream of Consciousness: Unity and continuity in conscious experience* New York, NY: Routledge.

—— 2004. The Self and the Phenomenal. *Ratio*, 17(4): 365–389.

—— 2010. Temporal Consciousness. In E. N. Zalta (ed), *The Stanford Encyclopedia of Philosophy* (*Fall Edition*).

Dehaene, S. 1997. *The Number Sense*. New York, NY: Oxford University Press.

Dennett, D. 1991. *Consciousness Explained*. Boston, MA: Little, Brown, and Co.

Dixon, N. and Spitz, L. 1980. The Detection of Auditory Visual Desynchrony. *Perception*, 9: 719–721.

Dresner, E. 2004. Over-Assignment of Structure. *Journal of Philosophical Logic*, 33: 467–480.

Dretske, F. 1981. *Knowledge and the Flow of Information*. Cambridge, MA: MIT Press.

Dunlap, J. C. and Loros, J. J. 2006. How Fungi Keep Time: Circadian System in *Neurospora* and Other Fungi. *Current Opinion in Microbiology* 9(6): 579–587.

Eagleman, D. and Holcombe, A. 2002. Causality and the Perception of Time. *Trends in Cognitive Science*, 6: 323–326.

Egan, F. 1995. Computation and Content. *The Philosophical Review*, 104: 181–203.

Emlen, S. T. 1975. Migration: Orientation and navigation. In D. S. Farner and J. R. King (eds), *Avian Biology*. New York, NY: Academic, 129–219.

Fleissner, Gerta, Schuchardt, K., Neumann, D., Bali, G., Falkenberg, G. and Fleissner, Guenther. 2008. A Lunar Clock Changes Shielding Pigment Transparency in Larval Ocelli of *Clunio marinus*. *Chronobiology International*, 25(1): 17–30.

Fodor, J. 2007. Revenge of the Given. In B. P. McLaughlin and J. Cohen (eds), *Contemporary Debates in the Philosophy of Mind*. New York, NY: Basil Blackwell pp. 105–116.

—— 2008. *The Language of Thought Revisited*. New York, NY: Oxford University Press.

Fujisaki, W., Shimojo, S., Kashino, M., and Nishida, S. 2004. Recalibration of Audiovisual Simultaneity. *Nature Neuroscience*, 7: 773–778.

Gallistel, C. R. 1990. *The Organization of Learning*. Cambridge, MA: MIT Press.

Gallistel, C. R., and Gelman, R. 1992. Verbal and preverbal counting and computation *Cognition*, 44: 43–74.

Gallistel, C. R. 2008. Learning and Representation. In R. Menzel and J. Byrne (eds), *Learning and Memory: A Comprehensive Reference* Vol. 1. New York, NY: Academic Press, pp. 227–242.

Gallistel, C. R. and King, A. P. 2009. *Memory and the Computational Brain: How Cognitive Science Will Transform Neuroscience*. New York: Blackwell/Wiley.
Geroch, R. 1978. *General Relativity: From A to B*. Chicago: University of Chicago Press.
Gibbon, J. 1977. Scalar expectancy theory and Weber's law in animal timing. *Psychological Review*, 84: 279–335.
Gibbon, J., Church, R. M., Meck, W. H. 1984. Scalar timing in memory. *Annals of The New York Academy of Sciences*, 423: 52–77.
Gibson, J. J. 1975. Events Are Perceivable but Time Is Not. In J. T. Fraser and N. Lawrence (eds.), *The Study of Time II. Proceedings of the Second Conference of the International Society for the Study of Time*. Berlin: Springer-Verlag, 295–301.
Giere, R. 1999. *Science Without Laws*. Chicago, IL: University of Chicago Press.
Gill, F. B. 1988. Trapline foraging by hermit hummingbirds: Competition for an undefended renewable resource. *Ecology*, 69: 1933–1942.
Golden, S. S. and Canales, S. R. 2003. Cyanobacterial circadian clocks-timing is everything. *Nature Reviews Microbiology*, 1:191–199.
Goldman, A. I. 1989. Metaphysics, Mind, and Mental Science. *Philosophical Topics*, 17: 131–145.
—— 1992. Epistemic Folkways and Scientific Epistemology. In A. I. Goldman, *Liaisons: Philosophy Meets the Cognitive and Social Sciences*. Cambridge, MA: MIT Press.
Goodman, N. 1951. *The Structure of Appearance*. Cambridge, MA: Harvard University Press.
—— 1976. *Languages of Art* Indianapolis, IN: Hackett.
Greco, J. 2010. *Achieving Knowledge: A Virtue-Theoretic Account*. New York, NY: Cambridge University Press.
Haugeland, J. 1981. Analog and analog. *Philosophical Topics*, 12: 213–26.
Heil, J. 2005. Dispositions. *Synthese*, 144: 343–356.
Helmholtz, H. v. 1873. The recent progress of the theory of vision. In Atkinson, E. (ed.) *Popular Lectures on Scientific Subjects*. New York, NY: D. Appleton and Company.
Hiatt, J. F. and Kripke, D. F. 1975. Ultradian Rhythms in Waking Gastric Activity. *Psychosomatic Medicine*, 37(4): 320–325.
Ivry, R. B. and Schlerf, J. E. 2008. Dedicated and intrinsic models of time perception. *Trends in Cognitive Science*, 12(7): 273–280.
James, W. 1890. *The Principles of Psychology I-II*. London, UK: Macmillan.
Kant, I. 2000. *Critique of Pure Reason*. New York, NY: Cambridge University Press.
Kaplan, D. (1989a) Demonstratives. In Almog, J., Perry, J. and Wettstein, H. (eds.) *Themes From Kaplan*. New York, NY: Oxford University Press.
—— 1989b. Afterthoughts. In Almog, J., Perry, J. and Wettstein, H. (eds.) *Themes From Kaplan*. New York, NY: Oxford University Press.
Karmarkar, U. R. and Buonomano, D. V. 2007. Timing in the absence of clocks: encoding time in neural network states. *Neuron*, 53: 427–438.
Killeen, P. 1975. On the temporal control of behavior. *Psychological Review*, 82: 89–115.
King, A. 2005. Multisensory Integration: Strategies for Synchronization. *Current Biology*, 15: 339–41.
Kripke, D. F. 1972. Perceptual Deprivation, REM Sleep, and an Ultradian Biological Rhythm. *Psychosomatic Medicine*, 34: 221–234.
—— 1973. Ultradian Rhythms in Sleep and Wakefulness. In E. D. Weitzman (ed), *Advances in Sleep Research* (New York: NY: Spectrum, 305–325).
Kripke, S. 1980. *Naming and Necessity*. Cambridge, MA: Harvard University Press.
Le Poidevin, R. 2004. A puzzle concerning time perception. *Synthese*, 142: 109–142.
—— 2007. *The Images of Time: An Essay on Temporal Representation*. Oxford: Oxford University Press.
Levine, J. 2010. Demonstrative Thought. *Mind and Language*, 25(2): 169–195.
Lewis, D. (1971) Analog and Digital. *Noûs*, 5: 321–328.
Libet, B. 2004. *Mind Time: The Temporal Factor in Consciousness*. Cambridge, MA: Harvard University Press.
Ludlow, P. 1999. *Semantics, Tense and Time*. Cambridge, MA: MIT Press.

Mach, E. 1959. *The Analysis of Sensations*. New York, NY: Dover Publications.

Malapani, C. and Fairhurst, S. 2002. Scalar timing in animals and humans. *Learning and Motivation*, 33: 156–176.

Matell, M. S. and Meck, W. H 2004. Cortico-striatal circuits and interval timing: Coincidence-detection of oscillatory processes. *Cognitive Brain Research*, 21: 139–170.

Matthews, R. J. 1994. The Measure of Mind. *Mind*, 103: 131–146.

—— 2007. *The Measure of Mind: Propositional Attitudes and Their Attribution*. New York, NY: Oxford University Press.

McGinn, C. 1989. *Mental Content*. Oxford, UK: Blackwell.

Meck, W. H. 2003. Introduction: The Persistence of Time. In W. H. Meck (ed), *Functional and Neural Mechanisms of Interval Timing* (*Methods and New Frontiers in Neuroscience*). Boca Raton, FL: CRC Press.

Meck, W. H. and Malapani, C. 2004. Introduction to Special Issue of the 'Neuroimaging of Interval Timing.' *Cognitive Brain Research*, 21: 133–137.

Mellor, D. H. 1981. *Real Time*. Cambridge, UK: Cambridge University Press.

—— 1998. *Real Time II*. New York, NY: Routledge.

Michener. G. R., 1979. The Circannual Cycle of Richardson's Ground Squirrels in Southern Alberta. *Journal of Mammalogy*, 60(4): 760–768.

Millar, A. 2003. A Suite of Photoreceptors Entrains the Plant Circadian Clock. *Journal of Biological Rhythms*, 18(3): 217–226.

Millikan, R. 1984. *Language, Thought, and Other Biological Categories*. Cambridge, MA: MIT Press.

—— 1993. On mentalese orthography. In Bo Dahlbom (ed.) *Dennett and his Critics*. Oxford Blackwell: 97–123.

Montemayor, C. and Balci F. 2007. Compositionality in Language and Arithmetic. *Journal of Theoretical and Philosophical Psychology*, 27(1): 53–72.

Montemayor, C. 2010. Time: Biological, Intentional and Cultural. In J. A. Parker, P. Harris, and C. Steineck (eds.), *Time: Limits and Constraints*. Leiden, The Netherlands: Brill.

Moore, C. 1990. Unpredictability and undecidability in dynamical systems. *Physical Review Letters*, 64: 2354–2357.

—— 1996. Recursion theory on the reals and continuous-time computation. *Theoretical Computer Science*, 162: 23–44.

Moore, R. Y. and Eichler, V. B. 1972. Loss of a circadian adrenal corticosterone rhythm following suprachiasmatic lesions in the rat. *Brain Research*, 42: 201–06.

Morsella, E. and Bargh, J. A. 2010. What is an output? *Psychological Inquiry*, 21 354–370.

Moyer, R. S. and Landauer, T. K. 1967. Time required for judgments of numerical inequality. *Nature*, 215: 1519–1520.

Mycka J. and Costa, J. F. 2007. A new conceptual framework for analog computation. *Theoretical Computer Science*, 347: 277–290.

O'Reilly, R. C. 2006. Biologically Based Computational Models of High-Level Cognition. *Science*, 314: 91.

Panda, S. and Hogenesch, J. 1998. It's All in the Timing: Many Clocks, Many Outputs. *Journal of Biological Rhythms*, 19(5): 374–387.

Paul, L. A. 2010. Temporal Experience. *Journal of Philosophy*, 107, 7: 333–359.

Peacocke, C. 1994. Content, Computation and Externalism. *Mind and Language*, 9(3): 303–335.

Penrose, R. 1996. *Shadows of the Mind: A Search for the Missing Science of Consciousness*. New York, NY: Oxford University Press.

Perry, J. 1979. The Problem of the Essential Indexical. *Nous*, 13: 3–20.

Pöppel, E. 1988. *Mindworks: Time and Conscious Experience*. New York, NY: Harcourt Brace Jovanovich.

Prior, A. N. 1959. Thank Goodness, That's Over. *Philosophy*, 34: 12–17.

—— 2003. *Papers on Time and Tense*. Oxford: Oxford University Press.

Pylyshyn, Z. W. 2007. *Things and Places: How the Mind Connects with the World*. Cambridge: MA: MIT Press.

Ralph, M. R., Foster, R. G., Davis, F. C. and Menaker, M. 1990. Transplanted suprachiasmatic nucleus determines circadian period. *Science* 247 (4945): 975–78.

Reichenbach, H. 1947. *Elements of Symbolic Logic*. New York, NY: Macmillan.

—— 1958. *The Philosophy of Space and Time*. New York, NY: Dover.

Renner, M. 1960. Contribution of the honey bee to the study of time sense and astronomical orientation. *Cold Spring Harbor Symposium on Quantitative Biology*, 25: 361–367.

Richter, C. P. 1967. Sleep and activity: Their relation to the 24 hour clock. In S. S. Kety, E. V. Evarts and H. L. Williams (eds.), *Sleep and altered states of consciousness*. Baltimore: Williams and Wilkins, pp. 8–29.

Roberts, W. 2007. Mental Time Travel: Animals Anticipate the Future. *Current Biology* 17 (11): 418–420.

Rosenbaum, D. A. 2002. Motor control. In H. Pashler (Series Ed.) & S. Yantis (Vol. Ed.), *Stevens' handbook of experimental psychology: Vol. 1. Sensation and perception*. New York, NY: Wiley. (3rd ed., pp. 315–339).

Russell, B. 1913. On the Experience of Time. In B. Russell, E. R. Eames (eds.) *The Collected Papers of Bertrand Russell: The 1913 Manuscript*, Vol. 7. Routledge, 1988.

—— 1937. *The Principles of Mathematics*. New York, NY: Norton.

Siegel, S. 2010. *The Contents of Visual Experience*. New York, NY: Oxford University Press.

Simen, P., Balci, F., deSouza, L., Holmes, P. and Cohen, J. D. 2011. A model of interval timing by neural integration. *Journal of Neuroscience*, 31 (25): 9238–9253.

Smart, J. J. C. 1980. Time and Becoming. In P. van Inwagen (ed.), *Time and Cause* (Dordrecht, Holland: D. Reidel).

—— 2006. Metaphysical Illusions. *Australasian Journal of Philosophy*, 84(2): 167–175.

Spence, C. and Squire, S. 2003. Multisensory Integration: Maintaining the Perception of Synchrony. *Current Biology*, 13: 519–521.

Stephan, F. K. and Zucker, I. 1972. Circadian rhythms in drinking behavior and locomotion activity of rats are eliminated by hypothalamic lesions. *Proceedings of the National Academy of Sciences* 69: 1583–1586.

Stetson, C., Cui, X., Montague, P. R. and Eagleman, D. M. 2006. Motor-Sensory Recalibration Leads to an Illusory Reversal of Action and Sensation. *Neuron*, 51: 651–659.

Stevens, S. S. 1959. Measurement, Psychophysics and Utility. In C. W. Churchman and P. Ratoosh (eds), *Measurement: Definitions and Theories*. New York, NY: John Wiley & Sons, pp. 18–63.

Stone, J. V., Hunkin, N. M., Porrill, J., Wood, R., Keeler, V., Beanland, M., Port, M. and Porter, N. R. 2001. When is Now? Perception of Simultaneity. *Proceedings of the Royal Society (B)*, 268: 31–38.

Stottinger, E., & Perner, J. 2006. Dissociating size representation for action and for conscious judgment: Grasping visual illusions without apparent obstacles. *Consciousness and Cognition: An International Journal, 15*, 269–284.

Strogatz, S. 2003. *Sync: The Emerging Science of Spontaneous Order*. New York, NY: Hyperion.

Suppes, P. and Zinnes, J. 1967. Basic Measurement Theory. In R. D. Luce (ed.), *Handbook of Mathematical Psychology*. New York, NY: Wiley and Sons.

Tye, M. 2003. *Consciousness and Persons: Unity and Identity*. Cambridge, MA: MIT Press.

van Fraassen, B. 1989. *Laws and Symmetry*. Oxford, UK: Clarendon Press.

Vander Wall, S. B. 1990. *Food hoarding in animals*. Chicago, IL: University of Chicago Press.

Velmans, M. 1991. Is human information processing conscious? *Behavioral and Brain Sciences*, 14(4): 651–668.

Wegner, D. M. 2002. *The Illusion of Conscious Will*. Cambridge, MA: MIT Press.

Wehner, R. and Müller, R., 1993. How Do Ants Acquire Their Celestial Ephemeris Function? *Naturwissenshaften* 80: 331–333.

Welch, R. B., DuttonHurt, L. D. and Warren, D. H. 1986. Contributions of Audition and Vision to Temporal Rate Perception. *Perception and Psychophysics*, 39: 294–300.

Wu, W. (2011) What is Conscious Attention? *Philosophy and Phenomenological Research*, 82(1): 93–120.

Xu, F., & Spelke, E.S. 2000. Large number discrimination in 6-month-old infants. *Cognition*, 74: 1–11.

Zimmerman, D. 2005. The A-theory of Time, the B-theory of Time, and 'Taking Tense Seriously.' *Dialectica*, 59(4): 401–457.

INDEX